"十二五"全国高校数字游戏设计专业精品教材

- 手游和虚拟现实方面最活跃、最易用的开发工具
- Unity的原生技术以及功能模块
- 案例剖析实际项目的功能构成与开发思路

Unity
游戏开发实用教程

万书帆 魏炜 王晖 邓兆静 编著

海洋出版社

2015年·北京

内 容 简 介

Unity 是由 Unity Technologies 开发的一款游戏开发引擎，也是目前手游和虚拟现实方面最活跃、最易用的开发工具。本书以循序渐进的方式，介绍了 Unity 的原生技术以及功能模块，同时通过丰富的案例剖析实际项目的功能构成与开发思路，使读者对 Unity 项目开发有一个清晰的思路。

全书共分为 10 章，着重介绍了 Unity 的基础知识、开发一个好的界面、多媒体应用、数据加载与卸载、Unity 读写外部数据、常用的组件、多人在线、基于 Unity 的安卓开发、常见问题错误及插件介绍等。最后通过制作多人在线的坦克大战游戏，介绍了使用 Unity 开发游戏的过程和方法。

在本书配套光盘中提供了本书游戏制作范例的工程文件、素材文件，方便读者学习参考。

适用范围：本书适合作为全国高校数字游戏设计专业教材、游戏制作培训班教材以及游戏设计师与爱好者的自学参考书。

图书在版编目(CIP)数据

Unity 游戏开发实用教程/万书帆等编著. —北京：海洋出版社，2015.3
ISBN 978-7-5027-9091-2

Ⅰ.①U… Ⅱ.①万… Ⅲ.①游戏程序－程序设计－教材 Ⅳ.①TP311.5

中国版本图书馆 CIP 数据核字（2015）第 034338 号

总　策　划：刘　斌	发行部：（010）62174379（传真）（010）62132549
责任编辑：刘　斌	（010）68038093（邮购）（010）62100077
责任校对：肖新民	网　　址：www.oceanpress.com.cn
责任印制：赵麟苏	承　　印：北京画中画印刷有限公司
排　　版：海洋计算机图书输出中心　晓阳	版　　次：2015 年 3 月第 1 版
	2015 年 3 月第 1 次印刷
出版发行：海洋出版社	开　　本：787mm×1092mm　1/16
地　　址：北京市海淀区大慧寺路 8 号（716 房间）	印　　张：10.25
100081	字　　数：246 千字
经　　销：新华书店	印　　数：1～4000 册
技术支持：（010）62100055	定　　价：45.00 元（含 1CD）

本书如有印、装质量问题可与发行部调换

前　　言

　　Unity 是由 Unity Technologies 开发的一款游戏开发引擎，也是目前手机游戏和虚拟仿真领域最活跃、最易用的开发工具。在近几年里，Unity 趁着移动平台的东风大肆扩张，现在 Unity 几乎成了整个游戏以及虚拟仿真领域的行业标准。为了方便用户熟悉 Unity 引擎，本书通过大量实用功能模块衍生的案例，详细地介绍了在 Unity 开发过程中涉及的主要功能点。

　　本书以循序渐进的方式，讲解了 Unity 原生技术以及功能模块，同时通过丰富的案例剖析实际项目的功能构成与开发思路，让读者在学完本书以后能对 Unity 项目开发有一个清晰的思路。本书所有涉及功能模块均是日常 Unity 开发过程中比较重要的知识点，在取材于实际项目的同时，进一步回归功能本身，让读者更加轻松的消化吸收。

　　本书内容分为 10 章，具体介绍如下。

　　第 1 章介绍关于 Unity 的一些基本知识和操作。

　　第 2 章介绍 Unity 界面系统中的控件以及一些综合案例，如小地图等。

　　第 3 章介绍音视频方面的应用。

　　第 4 章介绍资源文件的加载与卸载。主要介绍了两种加载资源方式：Reasources.load（）和 WWW 加载，并且简单介绍了内存管理的知识。

　　第 5 章介绍如何读写 Xml 和访问数据库。

　　第 6 章介绍自动寻路和地形创建。

　　第 7 章介绍多人在线方面的知识，实现多人在线聊天室和多人在线的动画同步。

　　第 8 章介绍关于 Unity 在安卓上的开发。本章制作了多个安卓方面的开发案例，方便读者熟悉 Unity 安卓开发。

　　第 9 章介绍一些 Unity 开发中可能遇到的问题以及 Unity 的一些插件。

　　第 10 章通过制作一个多人在线的坦克大战游戏，介绍 Unity 游戏开发的技巧和方法。

　　本书定位于 Unity 的初、中级读者，适合 Unity 爱好者和各行各业涉及使用此软件的人员作为参考书学习，同时也可作为职业院校以及计算机培训学校的教材。

　　本书由万书帆、魏炜、王晖、邓兆静编著，参加编写、校对、排版的人员还有胡斌、徐红香、万诗峰、张乐、蔡雯璐、黄舜尧、张静、李海洋、李川、郭佳伟、陈思敏、王旭、王海洋、张丹阳、张天杰、强彪、王波等。在此感谢购买本书的读者，虽然编者在编写本书的过程中倾注了大量心血，但恐百密之中仍有疏漏，恳请广大读者及专家不吝赐教。

　　最后，衷心希望您在本书的帮助下，能够熟练地掌握 Unity 引擎，设计并制作出优秀的 Unity 游戏！

编者

目　　录

第 1 章　Unity 基础应用 1
1.1　Unity 安装与卸载 1
1.1.1　Unity 安装 1
1.1.2　安装目录下部分文件夹简介 2
1.2　学习 Unity 的编辑界面 3
1.2.1　场景视图 3
1.2.2　游戏视图 5
1.2.3　Inspector 属性面板 6
1.2.4　层级（Hierarchy）面板 7
1.2.5　项目资源管理面板 7
1.3　Unity 资源导入与删除 7
1.3.1　模型等资源导入 7
1.3.2　加载与导出 unitypackage 10
1.4　Unity 的基本组件 14
1.4.1　摄影机 14
1.4.2　物理组件 16
1.4.3　灯光 17
1.4.4　寻路组件 18
1.4.5　音视频组件 18
1.4.6　网络组件 18
1.5　Mesh、Material 和 Texture 18
1.6　Unity 的一些自带脚本包 19
1.7　制作一个 Demo 19
1.7.1　Demo 的要求 19
1.7.2　搭建场景 19
1.7.3　建立目录并导入资源 22
1.7.4　建立脚本 26

第 2 章　开发一个好的界面 30
2.1　Unity 自带的界面系统 OnGUI 30
2.1.1　GUI.Label 30
2.1.2　GUI.Button 按钮 32
2.1.3　GUI.RepeatButton 长按状态按钮 33
2.1.4　GUI.DrawTexture 绘制纹理 34
2.1.5　GUI.Toggle 开关按钮 35
2.1.6　GUI.Toolbar 工具栏 36
2.1.7　GUI.TextField 单行文本输入框 37
2.1.8　GUI.TextArea 多行文本输入框 38
2.1.9　GUI.HorizontalSlider 水平滑动条 39
2.1.10　GUI.Window 窗口 40
2.1.11　GUIContent.Tooltip 工具提示 42
2.1.12　滚动视图 42
2.1.13　使用 unity 自带的控件实现一个树形列表 43
2.1.14　基于 OnGUI 下的屏幕自适应 45
2.1.15　制作一个简单的序列帧 46
2.1.16　制作一个简单的动态柱状图 46
2.1.17　制作一个图片查看器 48
2.1.18　制作一个小地图 52
2.2　NGUI 54
2.2.1　NGUI 概况 54
2.2.2　NGUI 与 OnGUI 的差别 54

第 3 章　多媒体应用 55
3.1　音频的控制 55
3.1.1　本地音频加载与播放 55
3.1.2　通过网络加载音频 58
3.2　视频播放控制 59
3.2.1　MovieTexture 的视频播放控制 59
3.2.2　AvPro QuickTime 的视频播放 62

第 4 章　数据加载与卸载 64
4.1　Resource.Load 加载资源 64
4.2　WWW 加载 66

第 5 章　Unity 读写外部数据 69
5.1　操作 Xml 69
5.1.1　C#操作 Xml 文件基础知识 69

5.1.2　Unity 加载 Xml 文件的
　　　　　方式 71
　　5.1.3　Unity 与 Xml 交互案例：用户
　　　　　登录验证 73
5.2　操作数据库 77
　　5.2.1　Xampp 介绍以及安装 77
　　5.2.2　在 Xampp 上建立一个数据库 79
　　5.2.3　创建一个 PHP 文件连接数据
　　　　　库 81
　　5.2.4　Unity+PHP+MySQL 操作数据
　　　　　库 82

第 6 章　常用的组件 85
6.1　导航网格 85
　　6.1.1　人物自动寻路到目标点 85
　　6.1.2　导航网格之 Off Mesh Link 使用 . 91
　　6.1.3　导航网格之动态障碍物
　　　　　Navmesh Obstacle 94
6.2　Terrain 地形系统 95

第 7 章　多人在线 101
7.1　开发一个多人聊天室 101
7.2　动画同步与位置同步 106

第 8 章　基于 Unity 的安卓开发 115
8.1　安卓开发环境配置 115
　　8.1.1　安装 jre 115
　　8.1.2　下载更新 android SDK 117
8.2　简单的触屏操作示例 121
　　8.2.1　单指旋转物体 121
　　8.2.2　多点缩放物体 122
8.3　在安卓上操作 Xml 123
　　8.3.1　安卓上如何读取 Xml 123
　　8.3.2　安卓上如何写入 Xml 124
8.4　安卓上播放视频 125

第 9 章　常见问题、错误及插件介绍 129
9.1　常见问题 129
　　9.1.1　js 脚本如何与 C#互相调用 129
　　9.1.2　Unity 脚本如何与网页脚本互
　　　　　相调用 131

　　9.1.3　Unity 发布为 Web 网页，在
　　　　　Web Player 中打开一个新页
　　　　　面不被拦截 133
　　9.1.4　如何打开一个摄像头 135
　　9.1.6　鼠标选中物体高亮 136
　　9.1.7　如何打开一个本地 EXE 138
9.2　常见错误及解决 139
　　9.2.1　在使用 Unity 编写脚本时遇到
　　　　　的错误 139
　　9.2.2　使用 WWW 崩溃如何解决 139
　　9.2.3　涉及 direct 3D11 特效有时候
　　　　　不能显示出效果 139
　　9.2.4　引用 dll 的时候报错 140
　　9.2.5　读取 Xml 错误 140
　　9.2.6　Fail to download data file 140
9.3　Unity 插件 141

第 10 章　多人在线的坦克大战 143
10.1　项目介绍 143
　　10.1.1　游戏主要功能描述 143
　　10.1.2　游戏开发步骤介绍 143
10.2　前期准备以及场景搭建 144
　　10.2.1　前期准备 144
　　10.2.2　搭建场景 144
　　10.2.3　设置游戏背景音乐 146
10.3　登录场景开发 147
　　10.3.1　登录场景界面制作 147
　　10.3.2　玩家注册功能 149
　　10.3.3　数据库登录验证 152
10.4　游戏场景开发 153
　　10.4.1　创建一个服务器 153
　　10.4.2　多人在线坦克行为模块开发 .. 154
　　10.4.3　登录后自动连接服务器并生
　　　　　　成玩家 155
　　10.4.4　炮弹的功能开发以及记分 ... 155
　　10.4.5　多人在线游戏小地图开发 ... 156
　　10.4.6　退出游戏并提交成绩到数据
　　　　　　库 157

第 1 章　Unity 基础应用

本章主要介绍 Unity 的基本知识，主要包括 Unity 的界面、组件以及创建脚本等。同时，介绍使用官方用户手册、脚本手册的方法。最后尝试写一个简单的 Unity 的 Demo，让大家感受 Unity 的开发方法。

1.1　Unity 安装与卸载

1.1.1　Unity 安装

登录 Unity 官方网站（http://Unity3d.com/Unity/download）下载 Unity 安装文件（本书开发环境为 Unity 4.3.2 及以上）。

下载完毕以后开始安装，双击 UnitySetup-4.3.2.exe，弹出如图 1-1 所示的界面，先单击"Next"，再单击"I Agree"，如图 1-2 所示。

图 1-1　安装界面一

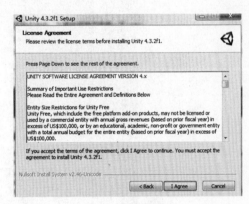

图 1-2　安装界面二

单击"Next"，选择一个安装路径，如图 1-3 所示。单击"Install"安装，如图 1-4 所示，显示安装进程，如图 1-5 所示。

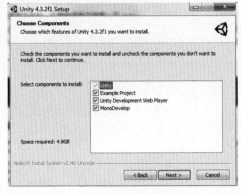

图 1-3　安装界面三

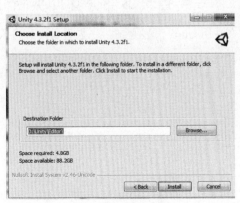

图 1-4　安装界面四

安装完成后，会显示如图 1-6 所示的界面，单击 "Finish"。

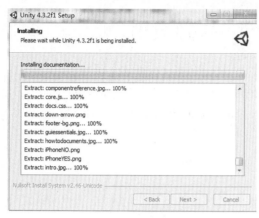

图 1-5　安装进程

图 1-6　安装完成图

安装完毕后需要激活 Unity，如图 1-7 所示。输入序列号激活即可；激活完成后就可以使用 Unity 进行开发了。

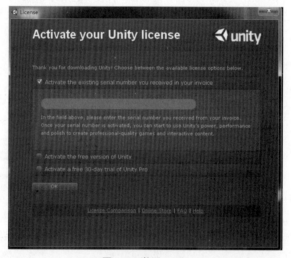

图 1-7　激活 Unity

1.1.2　安装目录下部分文件夹简介

在使用 Unity 开发之前，先来看看安装目录下面的文件夹。

1. Standard Packages

在 "\Unity\Editor\Standard Packages" 中，存放着一些资源包，如图 1-8 所示。

2. Lib

在 "\Unity\Editor\Data\Mono\lib" 下面存放着一些动态链接库，有了这些动态链接库，在开发时就不需要再到网上去找了，如图 1-9 所示。

图 1-8　自带的资源包

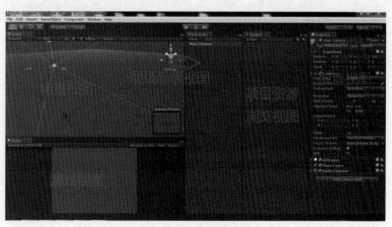

图 1-9　dll 动态库

1.2　学习 Unity 的编辑界面

安装完成以后，打开 Unity 就会看到如图 1-10 所示的编辑界面，它由几个标签窗口组成，称为"视图"。

图 1-10　Unity 编辑界面

1.2.1　场景视图

场景视图是 Unity 编辑游戏对象的场所。在这个场景里面可以直接通过鼠标来对游戏对象进行查看、选择、拖动、缩放、旋转，如图 1-11 所示。

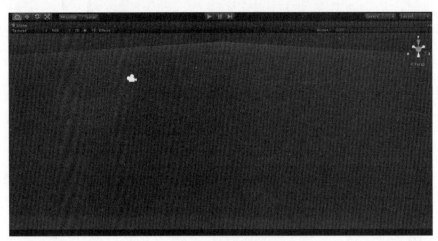

图 1-11　Scene 场景

如图 1-12 所示为一个创建了物体的场景。

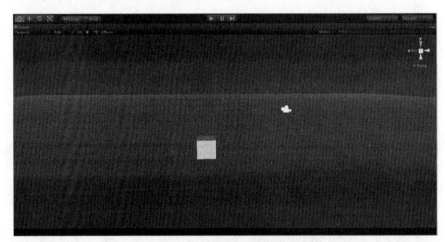

图 1-12　创建一个 Cube

单击鼠标左键可以选中物体；单击鼠标右键可以旋转视图并且鼠标的图标会变成一个"小眼睛"的形状；单击鼠标中键可以上下左右拖动视图并且鼠标的图标会变成一个"小手"的形状。滚动鼠标中键，可以拉近拉远视野与物体的距离。按 Alt+鼠标左键可以旋转整个视图；按 Alt+鼠标右键可以拉近拉远与物体的距离。

除了上面介绍的查看场景物体的方式以外，还可以通过同时按 W、A、S、D 键和鼠标右键来进入 Unity 飞行模式，全方位地查看整个场景。

场景视图的右上角是一个坐标轴，它可以显示场景相机的当前方向，并允许快速修改视图角度，如图 1-13 所示。

坐标轴的每个有色的轴臂表示一个几何轴。可以单击任意轴臂设置场景相机到该轴正交视图，还可以单击坐标轴下的文字来切换透视图和等距视图。在等距视图模式中，可以单击右键拖拽或单击 Alt 键+鼠标左键来拖拽摇移视图。

图 1-13

第 1 章　Unity 基础应用

在场景视图的右上方有 4 个按钮，单击手形按钮可以平移整个视图；第 2 个按钮是选择模式；单击第 3 个按钮可以旋转物体；单击第 4 个按钮可以缩放物体，如图 1-14 所示。

图 1-14　按钮

可以通过工具栏的平移、旋转和缩放工具分别操作游戏对象，如图 1-15 所示。

 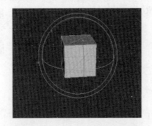 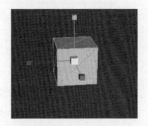

图 1-15　移动、旋转、缩放

第一个按钮是切换轴心点，第二个按钮是切换全局和局部坐标。

1.2.2　游戏视图

游戏视图是指运行后的预览视图。在游戏视图中可以看到游戏发布后运行的效果。一般在 Unity 中运行的效果和发布后的效果应该是一样的（当然有时候不一定，此时就需要好好调试和检查了），如图 1-16 所示就是游戏视图，也称作预览视图。

图 1-16　游戏视图

在 Standalone (1024×768) 中可以选择一个适合的窗口尺寸，可以在 File→Build Settings→Player Settings 下面的 Resolution and Presentation 中设置一个默认的分辨率，如 1024×768，表示窗口宽 1024 个像素×高 768 个像素。如图 1-17 所示。

图 1-17　设置分辨率

设置了 Player 以后单击 Standalone (1024×768) 就能找到设置的分辨率了。在开发的时候要尽量确定一个窗口大小，美工根据这个窗口尺寸来确定其他控件的大概位置以及图片的尺寸，同时在开发的时候还需要考虑屏幕的自适应，这里主要是界面的自适应，这部分内容会在介绍 OnGUI 的时候详细介绍。

在游戏窗口的 ![Maximize on Play Stats Gizmos] 中单击 Maximize on Play，表示在 Unity 编辑器运行游戏的时候最大化窗口；单击 Stats 会弹出一个显示渲染状态统计窗口，用于监控游戏的图形性能；单击 Gizmos 开关，所有显示在场景视图的 Gizmos 也将在游戏视图显示，包括所有使用 Gizmos 类函数绘制的 Gizmos。

下面主要介绍 Stats，它在开发中会经常使用，如图 1-18 所示。

- ![8.4 FPS (118.4ms)]：表示当前场景运行的帧率，一般在 60 以上就能流畅地运行。帧率尽量提高。
- ![Draw Calls: 33]：在 Unity 中，每次引擎准备数据并通知 GPU 的过程称为一次 Draw Call。Draw Call 对运行是否流畅有重要影响。如果不是迫不得已，应尽量降低 Draw Call。如果计算机配置为独显 512M 及以上，双核主频在 2.0 以上，发布为 PC 的应用在 1500 个 Draw Call 以下是能够接受的（其实还是有些卡）。如果是移动端这个数量级就难以接受了。至于到底多少 Draw Call 合适，应该根据发布的平台做针对性的测试。
- ![Saved by batching: 0]：如果用了遮挡剔除，那么会显示遮挡剔除减少的 Draw Call。
- ![Tris: 11.2k Verts: 14.1k]：Tris 表示在摄影机视野里渲染的三角面数总和；Verts 表示在摄影机视野里的顶点数目。这 2 个值是会变化的，它反映其实是摄影机实时渲染到的对象三角面数和顶点数之和。

图 1-18 Stats 窗口

1.2.3 Inspector 属性面板

Inspector 英文译为"检查、巡视"。在这个面板中展示的是选中的游戏对象的各种属性和信息。一个最基本的空物体包括名称、标签、所属层级、位置、旋转、缩放。其他游戏对象则可能包含更多的组件（脚本也可以看作是组件）。如图 1-19 所示是一个空物体的基本属性。

- ![Tag Untagged]：Tag 可以设置游戏物体的标签，默认为"Untagged"，可以单击 Tag 后面的下拉框添加自定义的 Tag。如图 1-20 所示。

图 1-19 空物体属性面板

图 1-20 设置 Tag 值

单击"Add Tag"会弹出如图 1-21 所示的界面，其中 Size 是指要添加标签的数量。Element 是内容。在程序中 Tag 可以通过 transform.tag 或者 gameObject.tag 来修改和访问。Tag 是一个很有用的属性。![Layer Default]（渲染层）也是一个非常实用的一个属性。单击 layer 后面的下拉菜单，选择 ![Add Layer...]，会弹出如图 1-22 所示的窗口。

图 1-21　添加 Tag 面板

图 1-22　Layer 添加面板

1.2.4　层级（Hierarchy）面板

层级面板中包括当前运行场景里面的游戏物体，在这个面板中一个游戏对象可以拥有它的子级对象。如图 1-23 所示，b 是 a 的父级，a 是 b 的子物体。

图 1-23　物体间的继承关系

某些对父物体的操作会影响子物体，如对父物体的位移、缩放、旋转，子物体都会受到影响。

1.2.5　项目资源管理面板

该面板中显示的是资源文件，它对应工程文件目录下 Assets 文件夹内的文件，也就是说只要将资源放到 Asset 文件夹下，Unity 就能够自动导入。如图 1-24、图 1-25 所示就是 Assets 对应的目录。

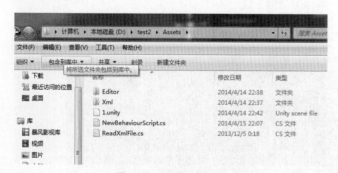
图 1-24　Assets 文件夹

图 1-25　Assets 文件夹下对应的 Project 面板

1.3　Unity 资源导入与删除

1.3.1　模型等资源导入

在开发之前需要将与工程相关的一些资源准备好并且导入到 Unity 中。将资源放到 Unity 工程目录下的 Assets 文件夹下，等到再次打开 Unity 的时候就能够自动导入了。例如，建立一个名为 Demo 的工程，如图 1-26、图 1-27 所示。

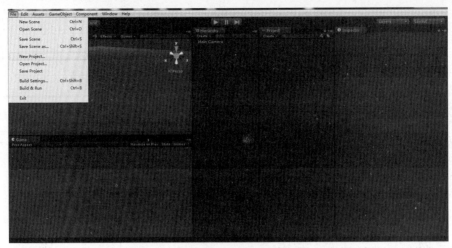

图 1-26　新建工程

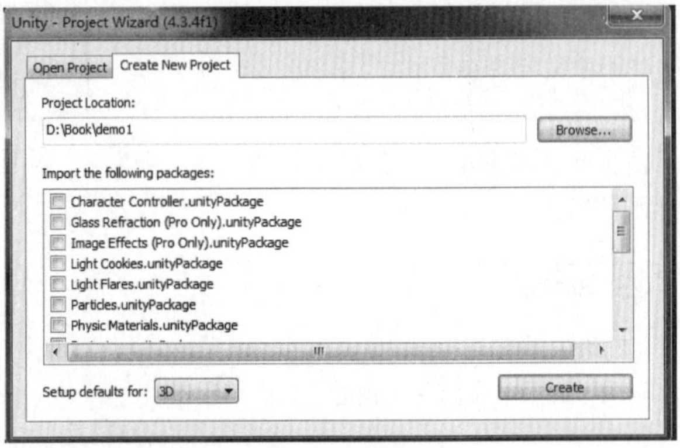

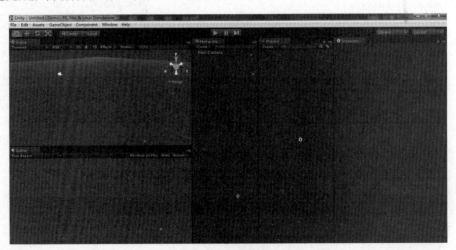

图 1-27　新建工程

创建完成后自动打开工程，如图 1-28 所示。

图 1-28　打开工程

可以看到这个场景里面没有任何物体，在 Hierarchy 下面只有一个 Main Camera（主摄影机）。下面将一些模型、图片、视频放到场景里，将鼠标放到 Project 上。

单击 Show in Explorer，如图 1-29 所示。双击 Assets 文件夹，将资源放到里面就可以了。如图 1-30 所示。

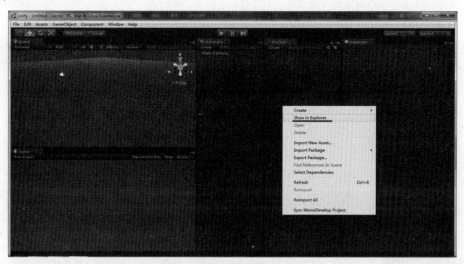

图 1-29　单击 Show in Explorer

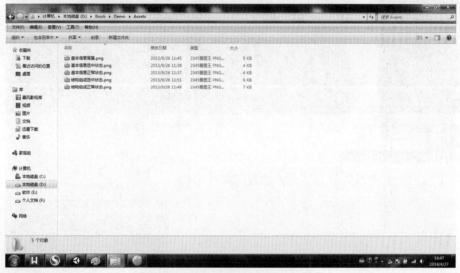

图 1-30　将图片放入 Assets 文件夹下

回到 Unity，资源就自动导入到里面了，如图 1-31 所示。

图 1-31　打开 Unity 自动导入

1.3.2 加载与导出 Unitypackage

加载与导出 Unitypackage 是一个很实用的 Unity 资源加载方式，特别是在要使用某个 Unity 的插件包来协助开发时。

继续使用 1.3.1 小节中的工程。在 Project 里面建一个文件夹，单击鼠标左键选择 Create->Folder，命名为 Images，将刚才几张图片都拖动到 Images 文件夹。也可以使用 Show in Explorer 后进入 Assets 文件夹下创建 Images 文件夹，将图片剪切到文件夹中。如图 1-32 所示。

单击 Images，按 F2 键可以重命名文件夹。

下面开始导出 Unitypackage。单击 Images，单击鼠标右键选择 Export Package，如图 1-33 所示。

图 1-32 创建文件夹放入图片

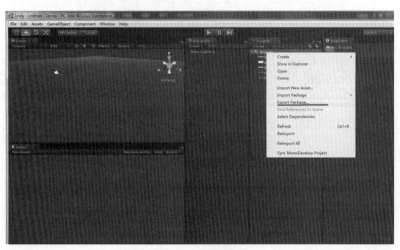

图 1-33 导出资源包

取消勾选 Include dependencies，单击 Export，将其命名为 image，导出到 E 盘根目录下，如图 1-34 所示为导出的文件。将弹出如图 1-35 所示的窗口。

图 1-34 导出资源包

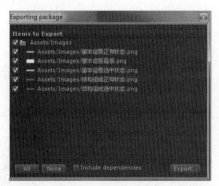

图 1-35 导出资源包

下面尝试导入这个资源包，注意不要在中文路径下导入 Unitypackage。

回到 Demo1，删除里面的所有文件。在删除 Project 面板中的资源文件时有 2 种方法：可以直接回到 Assets 目录下面删除，或者在 Unity 编辑软件下单击文件，然后单击右键选择 Delete，如图 1-36 所示。

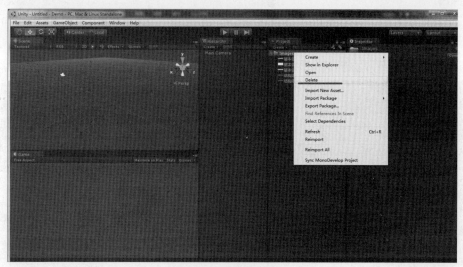

图 1-36 删除文件

删除资源后 Project 面板的显示如图 1-37 所示。

图 1-37 Project 面板

下面开始导入，直接单击 Project 面板，单击鼠标右键，弹出如图 1-38 所示界面。

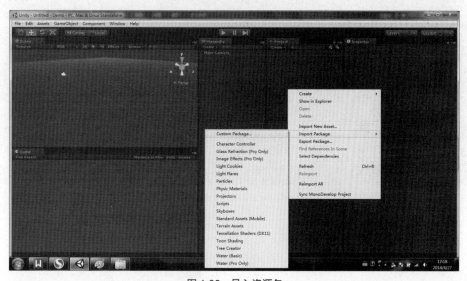

图 1-38 导入资源包

选择"Import Package→Custom Package",在弹出的窗口下选择 E 盘根目录下的 image.Unitypackage,如图 1-39 所示。

图 1-39 选择资源包导入

单击打开按钮,弹出如图 1-40 所示的窗口。

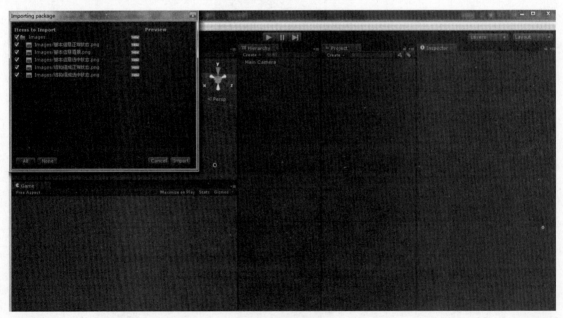

图 1-40 导入资源包

单击"Import",可以看到资源已经导入到场景中了,如图 1-41 所示。

上面导入的是自定义的资源包。Unity 为了方便用户开发提供了很多资源包,包括人物运动控制、水特效、图像特效、粒子特效等。下面导入人物控制的资源包,如图 1-42 所示。

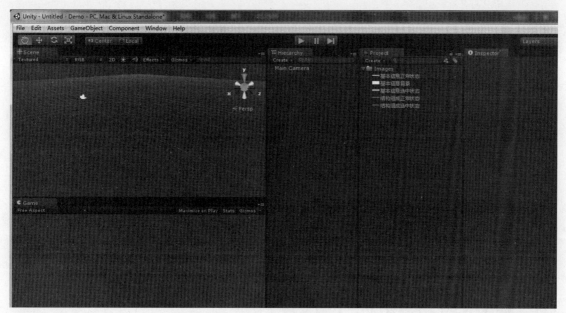

图 1-41 导入完成

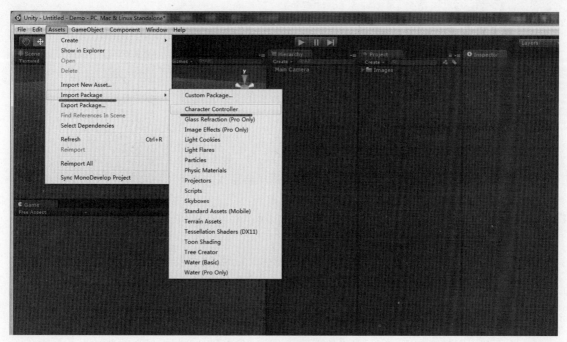

图 1-42 导入自带的资源包

单击"Import",如图 1-43 所示。在这个窗口中可以有选择性地导入资源,在每个资源文件前有个勾选项,勾选表示导入。有时候资源比较大,导入需要一些时间。导入后的 Project 显示的效果如图 1-44 所示。

最后,单击"File→Save Project"将工程保存即可。

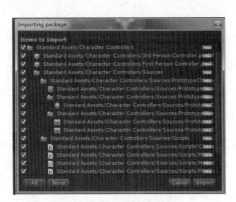

图 1-43 导入自带的资源包

图 1-44 导入后的人物运动控制资源包

1.4 Unity 的基本组件

下面介绍 Unity 的一些常用组件的属性含义。

1.4.1 摄影机

一个新建的工程场景里面必然有个主摄影机（Main Camera），如图 1-45 所示。

选择 Main Camera，按 F2 键可以重命名，摄影机的主要信息在 Inspector 面板下，如图 1-46 所示。

- Transform：主要包括摄影机的位置、旋转、缩放。在 Unity 中摄影机也是一个 GameObject。摄影机的缩放属性一般不使用。
- Camera：Camera 是摄影机重要属性，用于启用或关闭摄影机。
- Clear Flags：中文名为"清除标记"，每个摄影机在渲染场景的时候会存储颜色和深度信息。

图 1-45 主摄影机

- Clear Flags：其下面有 4 个选项：Skybox（天空盒）、Solid Color（纯色）、Depth Only（仅深度）、Don't Clear（不清除）。Skybox（天空盒）在后面第一个案例中会介绍；如果 Clear Flags 设置为 Solid Color，屏幕上的任何空的部分将显示当前相机的背景颜色；Depth-Only（仅深度）是一个非常重要的选项，如果场景里面有多个摄影机，深度低的摄影机渲染的画面在下面，摄影机渲染深度高，渲染的画面在上面。这就类似于画油画，先画背景，在背景之上又画一层，层层覆盖，形成画面。Depth Only 可以设置摄影机深度；Don't Clear 模式指不清除任何颜色或缓存，渲染的画面不会被下一帧清除。
- Background：是指摄影机背景颜色。
- Culling Mask：剔除遮罩，这个属性表示哪一个 Layer 可以被本摄影机渲染到屏幕。Everything 表示属于任何 Layer 层的物体都被本摄影机渲染；Nothing 表示任何一个 Layer 层的物体都不会被本摄影机渲染；每一个新建的物体默认是 Default 层。

关于 Layer 这里做一下解释：如果新建了一个物体或者导入了一个 Fbx 的文件到 Hierarchy 面板中，单击游戏物体，在 Inspector 面板查看一下属性，就会看到一个名为 Layer 的设置选项，如图 1-47 所示。

图 1-46 摄影机属性面板

图 1-47 Layer 为默认

单击 Layer 的设置选项，可以自己添加 Layer，如图 1-48 所示。

单击 Add Layer 会弹出如图 1-49 所示的窗口。

图 1-48 添加 Layer

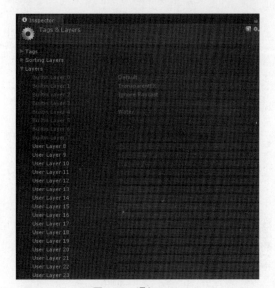

图 1-49 添加 Layer

可以从第 8 个 Layer 开始定义自己的 Layer。Unity 里面经常会用到 LayerMask 参数，它是一个 int 值。它的计算方法是这样的：例如"Builtin Layer +数值"或者"User Layer +数值"，它们的意思就是 2 的多少次方，也就是说 Layer 0 的 layerMask = 2 的 0 次方=1；Layer 1 的 LayerMask = 2 的 1 次方=2；Layer 2 的 LayerMask=2 的 2 次方=4 依次类推。如果想把前八层都要选中，那么 LayerMask 为多少呢？答案是这几个层的 LayerMask 相加或者用更高一级 Layer 的 LayerMask－1。

- `Projection` ：指投射，包括 Perspective 透视和 Orthographic 正交，当选中了正交摄影机时会有一个 Size 属性，可以设置正交摄影机的观察视野大小。

- `Field of View`：相机的视角宽度，可以在场景里面调试这个值，增加或减少会出现拉远拉近的效果。
- `Clipping Planes`：指裁剪平面，包括最近裁剪切面和最远裁剪切面，简单来说摄影机最近能看到的物体距离和最远能看到的距离。有时候减小最远裁剪切面的距离可以有效地提高游戏的性能。
- `Viewport Rect`：表示摄影机渲染的画面绘制在屏幕的位置，X、Y、W、H 分别表示起始点（X，Y）、长和高。
- `Rendering Path`：指该摄影机渲染使用的设置，Use Player Settings 使用播放器设置、Vertex Lit 顶点光照、Forward 快速渲染、Deferred Lighting 延迟照明；Target Texture 目标纹理可以将摄影机渲染的画面输出到渲染纹理（Render Texture）上。
- `HDR`：勾选它时将启用高动态范围图像，可以提供更多的动态范围和图像细节，能够更好地反映真实环境中的视觉效果。

1.4.2 物理组件

物理组件有很多，下面主要介绍几个常用的物理组件：盒碰撞器、网格碰撞器、角色控制器。

1. 盒碰撞器

盒碰撞器就是一个方形的碰撞器，通过"GameObject→Create Other→Cube"创建一个正方体，选中正方体后可以在 Inspector 面板下看到一个名为"Box Collider"的组件，如图 1-50 所示。

一个盒碰撞器包括 4 个属性：

- Is Trigger：如果勾选该项，那么就表明拥有该 Box Collider 的游戏物体是一个触发器，它将不受物理引擎控制，当碰撞的时候会发出触发信息。
- Material：需要是物理材质，它可以表现如何处理物体碰撞。
- Center：表示碰撞器的位置。
- Size：用来设置碰撞器的大小。

2. 网格碰撞器

Mesh Collider（网格碰撞器）比 Box Collider 多了 2 个设置项，当勾选 Smooth Sphere Collisions 后，碰撞检测使用平滑检测；如果激活 Convex，该网格碰撞器将会和其他网格碰撞器碰撞。如图 1-51 所示为网格碰撞器。

图 1-50　Box Collider

图 1-51　Mesh Collider

3. 角色控制器

当控制一个人物运动时会用到 Character Controller（角色控制器），它有以下几个设置选项：

- Height：角色控制器的高度。
- Radius：角色控制器的半径。
- Slope Limit：坡度限制，简单来说是指角色能爬多少角度的陡坡。
- Step Offset：控制角色可以通过多少高度以下的台阶。
- Min Move Distance：最小移动距离，如果角色移动的距离小于该值，那角色就不会移动。

这可以避免颤抖现象。大部分情况下该值被设为 0。
- Skin Width：皮肤厚度。皮肤厚度决定了两个碰撞器可以互相渗入的深度。较大的皮肤厚度值会导致颤抖，较小的皮肤厚度值会导致角色被卡住。一个合理的设定是使该值等于半径（Radius）的 10%。
- Center：表示角色控制器的位置。

1.4.3 灯光

Unity 中的灯光包括 Directional Light（方向光）、Point Light（点光源）、Spot Light（聚光）、Area Light（区域光）。单击"GameObject→Create Others"，可以看到灯光选项，如图 1-52 所示。

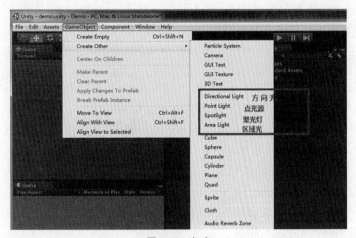

图 1-52　灯光

- Directional Light（方向光）：简单来说就是沿灯光照射方向变亮，逆方向为暗，整个场景都受影响。
- Point Light（点光源）：从它的位置向各个方向发出光线，影响其范围内的所有对象。
- Spot Light（聚光灯）：光线在按照聚光灯的角度和范围所定义的一个圆锥区域照射所有物体。
- Area Light（区域光）：只有在这个区域内的对象才会受到光线照射。

创建的灯光的设置面板如图 1-53 所示。

图 1-53　灯光设置选项

- Type：更改灯光的类型。
- Color：设置方向光的颜色。
- Intensity：设置灯光的强度。
- Cookie：这个纹理的 Alpha 通道作为一个遮罩，使光线在不同的地方有不同的亮度。如果灯光是聚光灯或方向光，这必须是一个 2D 纹理。如果灯光是一个点光源，它必须是一个立方图（Cubemap）。
- Cookie Size：缩放 Cookie 投影。
- Shadow Type：其下有三个选项，即无阴影、阴影、软阴影。
- Draw Halo：勾选后就有光晕。
- Flare：耀斑。

Unity 游戏开发实用教程

- Render Mode：其中 Atuo 表示在运行的时候根据附近灯光的亮度和当前的质量来确定。Important 表示灯光逐个像素渲染。Not Important 表示灯光总是以一种较快的方式进行渲染。
- Culling Mask：它决定哪些 Layer 层会受到灯光影响。
- Lightmapping：光照贴图。

1.4.4 寻路组件

导航网格是 Unity 提供给开发者开发寻路的一套组件，可以在"Component→Navigation"中找到相关组件，主要有以下三个：NavMesh Agent、OffMeshLink、Navmesh Obstacle。

- NavMesh Agent（导航网格代理组件）：用于如何寻路。
- OffMeshLink：用来将分离场景中的静态几何体的导航网格从一部分移动到另一部分。
- Navmesh Obstacle：用来设置动态障碍物，它可以被添加任何运动的游戏物体。

1.4.5 音视频组件

Unity 支持导入.aif、.wav、.mp3、.ogg 等音频文件格式和.xm、.mod、.it 和.s3m 等音轨模块。要想播放声音，需要给场景中某个游戏物体添加 Audio Source 组件，可以在"Component→Audio→Audio Source"中找到它。一般会把 Audio Source 添加到摄影机上。导入到 Unity 里面的声音文件，比如一个 MP3 文件会被 Unity 认为是一个声音剪辑（Audio Clip）。可以这样认识 Audio Source 和 Audio Clip 的关系：Audio Source 可以看作是一个音乐播放器，Audio Clip 就是一张 CD 专辑，CD 只有放到音乐播放器里面才能播放音乐，让人们听见。

Unity 提供了一个名为 MovieTexture 的类来操作视频，MovieTexture 继承于 Texture。通常 Unity 支持如下格式视频：.mov、.mpg、.mpeg、.mp4、.avi、.asf、.ogg。当一个视频文件添加到项目中，它会自动导入并转换为 Ogg Theora 格式。一旦影片纹理导入后，可以赋给一个游戏物体材质作为它的贴图。当影片纹理导入后，视频和声音是一起导入。影片纹理在 IOS 和安卓平台不支持。取而代之的是提供一个全屏幕的流媒体播放，可以使用 iPhoneUtils.PlayMovie 和 iPhoneUtils.PlayMovieURL 来播放视频。

1.4.6 网络组件

在开发多人在线游戏的时候，需要使用 Network View 这个组件。可以通过"Component→Miscellaneous→Network View"来对一个需要进行网络通信的物体添加网络组件。网络视图是网络共享数据的传输工具。通过网络视图可以发起两种类型的网络通信：状态同步和远程过程调用。

1.5 Mesh、Material 和 Texture

Unity 中的模型是由 Mesh、Material 与 Texture 三部分组成的，Mesh 网格可以看作是一个人的骨骼；Material 可以看作是人身上的肉；Texture 可以看作是人的皮肤。一个 Mesh 上可以有多个 Material，Material 可以被赋予不同的着色器（Shader）。在制作模型的时候最好能够统一模型、材质和贴图的名称，在对模型、材质、贴图命名时不要用中文。

18

1.6　Unity 的一些自带脚本包

下面介绍 Unity 自带的一些常用的脚本包，可以通过"Assets→Import Package+资源包来导入"，如图 1-54 所示。

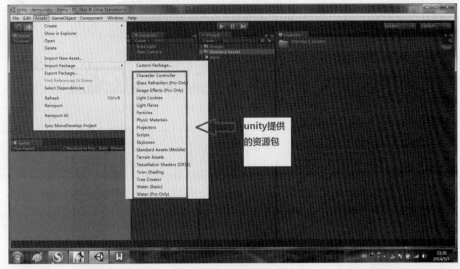

图 1-54　资源包

- Character Controller：提供了人物角色控制相关的资源与脚本，里面有一个简单的第一人称控制和第三人称控制方面的案例。
- Glass Refraction：提供了一套玻璃反射的材质和 Shader。
- Image Effect：包括很多关于图像特效方面的脚本。
- Particless：粒子特效包。
- Physic Materials：物理材质包。
- Sky Boxes：天空盒资源。
- Terrain Assets：地形资源包。
- Water：水特效。
- Tree Creator：用 Unity 创建的树的例子。

1.7　制作一个 Demo

1.7.1　Demo 的要求

这个小案例的主要功能如下：单击屏幕按钮进入场景；控制一个人物在场景内漫游；需要加入天空盒、灯光；发布为 EXE；添加一个背景音乐；单击 ESC 能退出应用。

1.7.2　搭建场景

双击打开 Unity，进入 Unity 后单击"File→New Project"，如图 1-55 所示，在弹出窗口中选择要新建的工程文件路径，一般使用英文路径且工程名为英文。这里创建一个名为"Hello_Unity_demo"的工程文件，这里不勾选其他资源包，然后保存路径，如图 1-56 所示。

图 1-55　创建工程文件

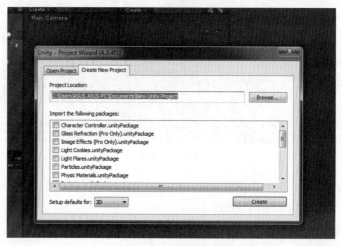

图 1-56　保存路径

创建完成后如图 1-57 所示。

图 1-57　创建的场景

如果打开的界面布局不同，可以单击右上角的下拉列表"Layout →2 by 3"，如图 1-58 所示。

然后修改 Project 面板布局，如图 1-59 所示。

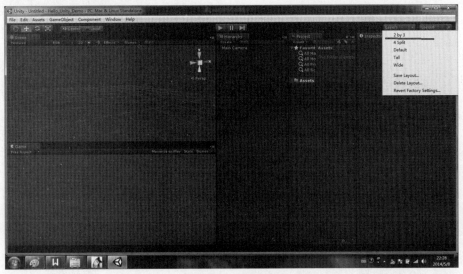

图 1-58　界面布局

如果想把当前这种布局方式保存下来，可以单击"Layout→Save Layout"，在弹出的窗口中修改名称为"MyLayout"。设置过程如图 1-60 所示，单击"Save"即可，如图 1-61 所示。以后如果想用自己保存的这个布局，可以在下拉菜单里找到，如图 1-62 所示。

图 1-59　修改面板布局

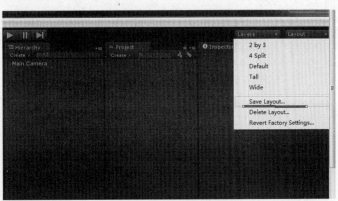

图 1-60　单击 Save Layout

图 1-61　保存布局

图 1-62　保存过的布局

1.7.3 建立目录并导入资源

在 Project 面板下建立"Scripts"、"Scenes"、"Models",如图 1-63 所示,鼠标右键单击 Project 面板中的"Create → Folder";也可以直接找到工程文件夹,在工程文件 Assets 中建立这 3 个文件夹,如图 1-64 所示。

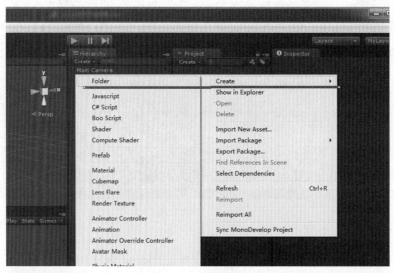

图 1-63 建立目录

图 1-64 建立目录

将 Fbx 文件复制到"Hello_Unity_Demo"工程中的 Models 下,如图 1-65、图 1-66 所示。

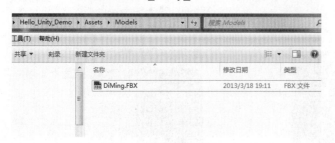

图 1-65 选择 Fbx 文件

图 1-66 复制文件

在 Scripts 下面建立两个 C#脚本,分别名为 MainProcess 和 MainUI,如图 1-67、图 1-68 所示。

按"Ctrl+S"键保存场景到 Scenes 文件夹下并命名为"Demo",如图 1-69 所示。

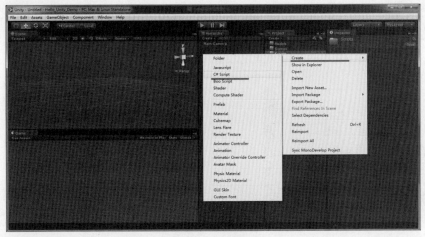

图 1-67　选择命令　　　　　　　　　　　　　图 1-68　建立脚本

将 Models 下的 Fbx 文件拖入到 Hierarchy 面板中，如图 1-70 所示。

图 1-69　保存 Demo　　　　　　　　　图 1-70　将 Fbx 文件拖入到 Hierarchy 面板中

为场景内的物体添加碰撞，注意是对有 Mesh 的物体添加，可以全选模型下的子物体，然后单击"Component→Physics→Mesh Collider"，如图 1-71 所示。

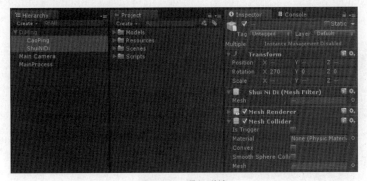

图 1-71　添加碰撞

下面给场景打上灯光，添加一盏方向光并且有阴影效果，单击"GameObject → Create Other → Directional Light"，如图 1-72 所示。方向光的参数如图 1-73 所示。

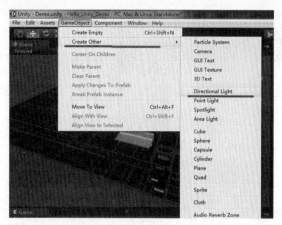

图 1-72 添加灯光

图 1-73 方向光灯光设置

下面为场景添加天空盒。单击"Assets→Import Package→Skyboxes",如图 1-74 所示。加载完成后在 Project 面板下会有一个名为"Skyboxes"的文件夹,里面有一些天空盒材质,如图 1-75 所示。

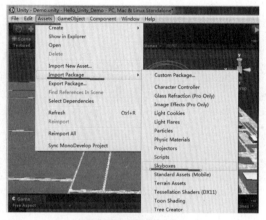

图 1-74 添加天空盒

图 1-75 导入的天空盒资源包

单击"Edit→Render Settings"进行设置,然后拖动一个材质球到 RenderSettings 面板下的 Skybox Material 上,如图 1-76 所示。

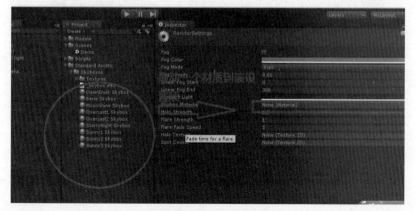

图 1-76 设置天空盒材质

这样天空盒就设置完成了，让我们看看效果吧！如图1-77所示。

将人物导入到场景中。单击"Assets→Import Package→Character Controllers"，导入完成后在Project面板中会看到Charater Controllers文件夹，如图1-78所示。

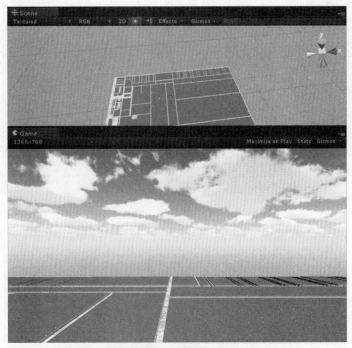

图1-77 添加天空盒

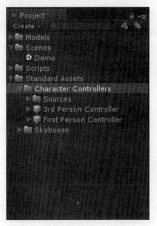

图1-78 导入自带的人物资源包

将Sources文件夹下的"3rd Person Controller"模型文件拖到Hierarchy面板下，然后设置人物身上的动画，如图1-79所示。也可以直接在Project中将动画拖到设置面板上，如图1-80所示。

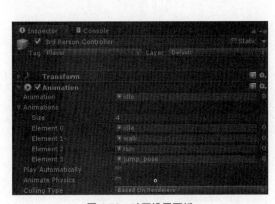

图1-79 动画设置面板

图1-80 Project面板下的动画

设置Third Person Controller.cs脚本，如图1-81所示。最后设置Third Person Camera.cs，如图1-82所示。

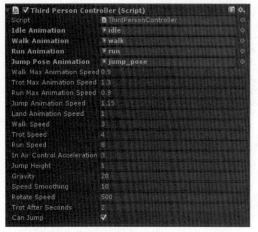

图 1-81 设置 Third Person Controller.cs

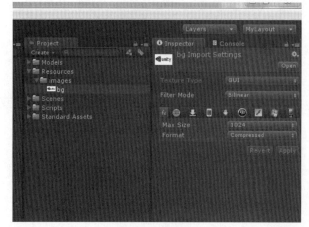

图 1-82 设置 Third Person Camera.cs

1.7.4 建立脚本

下面为这个 Demo 添加一个简单的登录界面和单击 Esc 退出。

在 Unity 工程中建立一个 Resources 文件夹，注意这个文件夹的文件名不要更改，否则在使用 Resources.Load()这个函数的时候会有问题。在 Resources 文件夹下建立一个 Images 文件夹。在 Images 下有张图片名为"bg"，图片设置如图 1-83 所示。

双击打开 Scripts 下面的 MainUI.cs 脚本，开始编写界面部分代码。关于界面的代码需要写在 OnGUI 这个函数中，注意不能写在 Update 这个函数中。代码如下：

图 1-83 设置图片

```
using UnityEngine;
using System.Collections;

public class MainUI : MonoBehaviour {

    // 背景
    private Texture2D loadBg;
    // 登录按钮风格
    private GUIStyle loadButtonStyle;

    private bool isLoad = false;
    void Start ()
    {
        // 从 Resources 文件夹下导入 bg 图片（注意不要有后缀名）
        loadBg = Resources.Load ("images/bg") as Texture2D;
    }
```

```
void OnGUI ()
{
    if(!isLoad)
    {
        GUI.DrawTexture( new Rect (0,0,Screen.width,Screen.height),loadBg);
        if (GUI.Button (new Rect (Screen.width / 2, Screen.height / 2, 100, 60), "Load...."))
        {
            isLoad = true;
        }
    }
}
```

将 MainUI 拖到 MainProcess 这个游戏物体上，然后单击运行，完成效果如图 1-84 所示。

下面添加一个退出控制，单击进入 Main Process.cs 脚本，添加如下一段代码：

```
//当 Esc 被按下退出
    if(Input.GetKeyDown(KeyCode.Escape))
    {
        Application.Quit();
    }
```

这段代码要放到 Update()函数中。

最后为这个应用添加一段背景音乐，默认循环播放。找到在 Player 下面的 Main Camera 游戏物体，为该游戏物体添加 Audio Source 组件。在 Resource 下面建立一个名为"Audio"的文件夹，放入一个 mp3 文件，如图 1-85 所示。

图 1-84　效果图

图 1-85　背景音乐

将这个音乐文件拖到"Main Camera→Audio Source→Clip"中，如图 1-86 所示。注意要有 Audio Listener 这个组件，如果没有的话，就无法听到声音。

最后将这个 Demo 发布为 EXE。单击"File→Build Settings"，在弹出的窗口里单击"Add Current"，如图 1-87 所示为发布过程。

发布完成后可以看到发布路径下有 2 个文件，如图 1-88 所示。

单击 EXE 文件就可以打开刚才制作的应用了，如图 1-89 所示，打开后可以设置分辨率和画面质量。

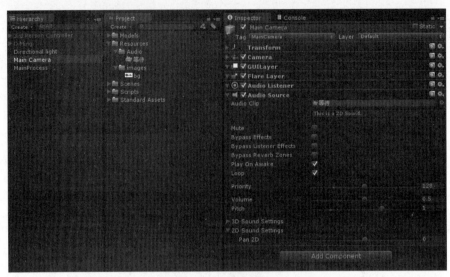

图 1-86　添加背景音乐

图 1-87　发布设置

图 1-88　发布 EXE

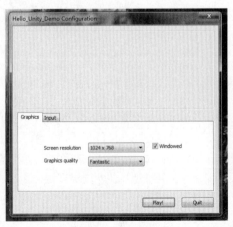

图 1-89　打开设置

如果勾选 Windowed，那么就是窗口模式，否则为全屏。单击 Play 可以运行应用，单击 Quit 退出，如图 1-90 所示为打开后的效果，在场景中单击 ESC 可以退出。

图 1-90　完成效果

第 2 章　开发一个好的界面

本章主要介绍 OnGUI 和第三方界面插件 NGUI 这两种界面系统的控件使用，并配合一些实际的界面案例，介绍两种界面系统的差异。

2.1　Unity 自带的界面系统 OnGUI

在开始界面开发前，先新建一个名为"HelloGUI"的工程，并且在这个工程里创建一个控制界面的脚本 UIManager.cs，如图 2-1 所示。

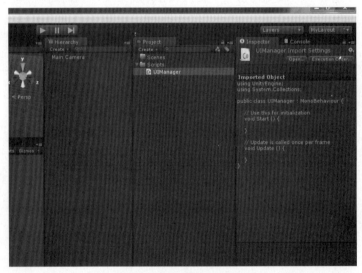

图 2-1　创建 UIManager.cs

2.1.1　GUI.Label

单击打开 UIManager.cs 脚本后，删除 Update()函数，改为 void OnGUI()。Unity 自带的界面函数必须要在 OnGUI 这个函数中运行。下面介绍 OnGUI 中常用的控件以及控件函数。

- Rect：由 X、Y 位置和 Width、Height 大小定义的二维矩形。Rect 结构主要用于 2D 操作。
- GUI.Label：其中的常用函数如下：

```
Static function Label (position : Rect, text : string) : void
    static function Label (position : Rect, image : Texture) : void
    static function Label (position : Rect, text : string, style : GUIStyle) : void
    static function Label (position : Rect, image : Texture, style : GUIStyle) : void
```

- Position：位置大小。
- Text：文字内容。
- Image：图片。
- Style：标签的风格，比如文字大小、字体、颜色等。

下面创建一个界面，效果如图 2-2 所示。工程目录如图 2-3 所示。

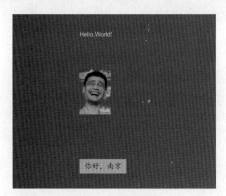

图 2-2　Label 示例　　　　　　　　　图 2-3　工程文件目录

源代码如下：

```
using UnityEngine;
using System.Collections;

public class UIManager : MonoBehaviour
{
    private Texture2D tex;
    private GUIStyle labelStyle;
    private GUIStyle labelStyle2;
    private Font myFont;
    void Start ()
    {
        // 从默认 Resources 文件夹下导入字体
        myFont = Resources.Load("Font/simkai") as Font;
        // 从 Resources 文件夹下的 Textures 下导入一张名为 tex 的图片
        tex = Resources.Load("Textures/tex") as Texture2D;

        // 创建一个界面风格，设置它的字体为楷体，字体颜色为红色，字号为 35
        labelStyle = new GUIStyle ();
        labelStyle.font = myFont;
        labelStyle.normal.textColor = Color.red;
        labelStyle.fontSize = 35;
// 创建一个界面风格，设置字体为楷体，用 labelBg 这张图作为通常状态下的背景，文字居中对齐
        labelStyle2 = new GUIStyle();
        labelStyle2.font = myFont;
        labelStyle2.normal.background = Resources.Load("Textures/labelBg") as Texture2D;
        labelStyle2.alignment = TextAnchor.MiddleCenter;


    }

    void OnGUI ()
    {
            //1.文字标签
        GUI.Label (new Rect (300, 100, 200, 50), "Hello,World!");
```

```
            // 2.图片
            GUI.Label (new Rect (300, 180, 100, 100), tex);
                   // 设置了文字效果的标签
            GUI.Label (new Rect (300, 300, 100, 50), "你好，南京！",labelStyle);
                   // 带背景的标签
            GUI.Label(new Rect(300,370,100,30),"你好，南京",labelStyle2);


    }
}
```

2.1.2 GUI.Button 按钮

常用的几个函数如下：

```
static function Button (position : Rect, text : String) : bool
static function Button (position : Rect, image : Texture) : bool
static function Button (position : Rect, text : String, style : GUIStyle) : bool
```

参数解释：
- Position：按钮在屏幕上的位置。
- Text：在按钮上显示的文本。
- Image：按钮上显示的图片。
- Style：风格，如果不使用，按钮风格应用当前的 GUISkin 皮肤。

下面创建一个按钮，效果如图 2-4 所示。

源代码如下：

图 2-4 按钮示例

```
using UnityEngine;
using System.Collections;

public class UI_Button : MonoBehaviour
{
    private GUIStyle buttonStyle;
    private Font myFont;
    private Texture2D tex;
    void Start ()
    {
        // 导入字体
        myFont = Resources.Load("Font/simkai") as Font;
        /*创建按钮风格，按钮主要有三种状态：按下效果，正常效果，碰到悬停效果。*/
        buttonStyle = new GUIStyle();
        buttonStyle.normal.background =    buttonStyle.active.background =
Resources.Load("Textures/Button/normal") as Texture2D;
        buttonStyle.hover.background = Resources.Load("Textures/Button/hover") as Texture2D;
        buttonStyle.font = myFont;
        /*设置了对齐方式为居中，字号 20*/
        buttonStyle.alignment = TextAnchor.MiddleCenter;
        buttonStyle.fontSize = 20;
```

```
            tex = Resources.Load("Textures/Button/hover") as Texture2D;

        }

        void OnGUI ()
        {
            if(GUI.Button(new Rect(100,100,100,50),"Hello Button"))
            {
                print("默认使用当前样式");
            }

            if(GUI.Button(new Rect(100,200,150,50),tex))
            {
                print("带图片的按钮");
            }

            if(GUI.Button(new Rect(100,300,150,50),"你好，南京！ ",buttonStyle))
            {
                print("使用当前样式的按钮");
            }
        }
}
```

2.1.3 GUI.RepeatButton 长按状态按钮

常用的几个函数如下：

```
static function RepeatButton (position : Rect, text : String) : bool
    static function RepeatButton (position : Rect, image : Texture) : bool
    static function RepeatButton (position : Rect, text : String, style : GUIStyle) : bool
```

RepeatButton 参数和上面的控件参数类似，RepeatButton 与一般 Button 的不同在于 RepeatButton 表示的是按钮一直按着，这是一个持续的长状态；Button 单击的时间是一个短暂的状态。

下面以第三个函数为例，创建一个长按状态按钮，完成效果如图 2-5 所示。

源代码如下：

图 2-5 RepeatButton 示例

```
using UnityEngine;
using System.Collections;

public class UI_RepeatButton : MonoBehaviour {

    private GUIStyle buttonStyle;
    private Font myFont;
    void Start ()
    {
```

```
        // 导入字体
        myFont = Resources.Load("Font/simkai") as Font;
        /*创建按钮风格,按钮主要有三种状态：按下效果，正常效果，碰到悬停效果，*/
        buttonStyle = new GUIStyle();
        buttonStyle.normal.background = buttonStyle.onNormal.background = Resources.Load("Textures/Button/normal") as Texture2D;
        buttonStyle.hover.background = Resources.Load("Textures/Button/hover") as Texture2D;
        buttonStyle.font = myFont;
        /*设置了对齐方式为居中，字号 20*/
        buttonStyle.alignment = TextAnchor.MiddleCenter;
        buttonStyle.fontSize = 20;

    }


    void OnGUI ()
    {
        if(GUI.RepeatButton(new Rect(500,200,150,50),"你好，南京！",buttonStyle))
        {
            GUI.Label(new Rect(500,300,300,50),"一直按着按钮，出现标签");
        }
    }
}
```

2.1.4 GUI.DrawTexture 绘制纹理

常用的函数如下：

static function DrawTexture (position : Rect, image : Texture）: void

这个函数有两个参数，第一个确定图片的位置和长宽；第二个参数是需要绘制的贴图。

下面绘制纹理效果，如图 2-6 所示。

图 2-6　DrawTexture 示例

源代码如下：

```
using UnityEngine;
using System.Collections;

public class UI_DrawTexture : MonoBehaviour
{

    private Texture2D tex;
    void Start ()
    {
        // 导入图片
        tex = Resources.Load("Textures/DrawTexture/Flower") as Texture2D;
    }


    void OnGUI ()
    {
        GUI.DrawTexture(new Rect(0, 0, 1366, 768),tex);
    }
}
```

2.1.5 GUI.Toggle 开关按钮

常用的函数如下：

```
static function Toggle (position : Rect, value : bool, text : String) : bool
        static function Toggle (position : Rect, value : bool, text : String, style : GUIStyle) : bool
```

参数说明：
- Position：控件在屏幕位置的宽和高。
- Value：设置开关。
- Text ：按钮上显示的文字。
- Style：开关样式。

下面制作一个开关按钮，效果如图 2-7 所示。

源代码如下：

图 2-7　Toggle 示例

```
using UnityEngine;
using System.Collections;

public class UI_Toggle : MonoBehaviour
{

    private bool isStudent = false;
    private bool isTeacher = false;
    private bool isOpen = true;
    private GUIStyle toggleStyle;
    void Start ()
    {
        toggleStyle = new GUIStyle();
```

```
            // 没有按下的时候状态
            toggleStyle.normal.background = Resources.Load("Textures/Toggle/关闭") as Texture2D;
            // 按下的时候状态
            toggleStyle.onNormal.background = Resources.Load("Textures/Toggle/打开") as Texture2D;
    }


    void OnGUI ()
    {
            isStudent = GUI.Toggle(new Rect(300,200,100,50),isStudent,"Student");
            isTeacher = GUI.Toggle(new Rect(300,250,100,50),isTeacher,"Teacher");

            isOpen = GUI.Toggle(new Rect(300,300,250,250),isOpen,"",toggleStyle);
    }
}
```

2.1.6　GUI.Toolbar 工具栏

常用的函数如下：

```
static function Toolbar (position : Rect, selected : int, texts : String[]) : int
static function Toolbar (position : Rect, selected : int, texts : String[], style : GUIStyle) : int
```

参数说明：

- Position：用于工具栏在屏幕上的矩形位置。
- Selected：被选择按钮的索引号。
- Texts：显示在工具栏按钮上的字符串数组。
- Style：风格。

下面制作一个工具栏，效果如图 2-8 所示。

源代码如下：

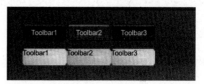

图 2-8　Toolbar 示例

```
using UnityEngine;
using System.Collections;

public class UI_Toolbar : MonoBehaviour
{

    private int toolbarInt = 0;
    private int toolbarInt1 = 0;
    private string[] toolbarStrings = new string[] {"Toolbar1", "Toolbar2", "Toolbar3"};
    private GUIStyle toolbarStyle;
    void Start()
    {
            /*这里需要注意的是 Toolbar 设置的风格主要有效的是没有按下和按下后这两种状态，按下这个状
态是一个持续的状态*/
            toolbarStyle = new GUIStyle();
            toolbarStyle.normal.background = toolbarStyle.active.background = Resources.Load("Textures/Button/normal") as Texture2D;
            //toolbarStyle.hover.background = Resources.Load("Textures/Button/hover") as Texture2D;
            toolbarStyle.onNormal.background = Resources.Load("Textures/Button/hover") as Texture2D;
```

```
        }
        void OnGUI()
        {
            toolbarInt = GUI.Toolbar(new Rect(25, 25, 250, 30), toolbarInt, toolbarStrings);

            toolbarInt1 = GUI.Toolbar(new Rect(25, 70, 250, 30), toolbarInt1, toolbarStrings,toolbarStyle);
        }
}
```

2.1.7 GUI.TextField 单行文本输入框

常用的函数如下：

```
static function TextField (position : Rect, text : String) : String
    static function TextField (position : Rect, text : String, maxLength : int) : String
    static function TextField (position : Rect, text : String, style : GUIStyle) : String
    static function TextField (position : Rect, text : String, maxLength : int, style : GUIStyle) : String
```

参数说明：

- Position：控件在二维屏幕上的位置和宽高。
- Text：输入框内的文字。
- MaxLength：文本框中最大字符长度。
- Style：控件风格样式。

下面制作一个单行文本输入框，如图 2-9 所示。

源代码如下：

图 2-9 TextField 示例

```
using UnityEngine;
using System.Collections;
public class UI_TextField : MonoBehaviour
{

    private string stringToEdit1;
    private string stringToEdit2;
    private string stringToEdit3;
    private GUIStyle textFieldStyle;

private Texture2D textFieldBg;
    void Start ()
    {
        stringToEdit1 = "";
        stringToEdit2 = "";
        stringToEdit3 = "";

        textFieldStyle = new GUIStyle();
        // 使用中文楷体
        textFieldStyle.font = Resources.Load("Font/simkai") as Font;
        // 采用居中左对齐方式
        textFieldStyle.alignment = TextAnchor.MiddleLeft;

        textFieldBg = Resources.Load("Textures/TextField/输入框") as Texture2D;
```

```
                textFieldStyle.normal.background = textFieldBg;
                textFieldStyle.fontSize = 30;
        }
void OnGUI ()
{
        // 使用默认皮肤效果的文本输入框
        stringToEdit1 = GUI.TextField(new Rect(400,110,300,50),stringToEdit1);
        // 使用自己定义的样式
stringToEdit2 = GUI.TextField(new Rect(400,170,300,50),stringToEdit2,textFieldStyle);
// 限制输入框最多能输入 10 个字符
stringToEdit3 = GUI.TextField (new Rect(400,230,300, 50),stringToEdit3,10,textFieldStyle);
}
}
```

2.1.8　GUI.TextArea 多行文本输入框

常用的函数如下：

```
static function TextArea (position : Rect, text : String) : String
    static function TextArea (position : Rect, text : String, maxLength : int) : String
    static function TextArea (position : Rect, text : String, style : GUIStyle) : String
    static function TextArea (position : Rect, text : String, maxLength : int, style : GUIStyle) : String
```

参数说明：
- Position：用于文本区域在屏幕上矩形的位置。
- MaxLength：控制字符串的最大长度，如果不设置，用户可以一直输入。
- Style：样式。

下面制作一个多行文本输入框，效果如图 2-10 所示。

源代码如下：

图 2-10　TextArea 示例

```
using UnityEngine;
using System.Collections;

public class UI_TextArea : MonoBehaviour {

    private string stringToEdit;
    private GUIStyle TextAreaStyle;
        private string stringToEdit1;
        void Start ()
        {
            stringToEdit = "Hello World\nHello NanJing";
            stringToEdit1 ="快乐中国！\n 欢乐南京";
            TextAreaStyle = new GUIStyle();
            TextAreaStyle.normal.background = Resources.Load("Textures/TextArea/1") as Texture2D;
            TextAreaStyle.alignment = TextAnchor.MiddleCenter;
            TextAreaStyle.font = Resources.Load("Font/simkai") as Font;


        }
```

```
void OnGUI ()
{
    // 最多输入 200 个字符
    stringToEdit = GUI.TextArea(new Rect(10, 10, 200, 100), stringToEdit, 200);

        stringToEdit1 = GUI.TextArea(new Rect(10, 120, 200, 100), stringToEdit1, 200,TextAreaStyle);
    }
}
```

2.1.9 GUI.HorizontalSlider 水平滑动条

常用的函数如下：

```
static function HorizontalSlider (position : Rect, value : float, leftValue : float, rightValue :    float) : float
        static function HorizontalSlider (position : Rect, value : float, leftValue : float, rightValue :    float, slider : GUIStyle, thumb : GUIStyle) : float
```

参数说明：

- Position：控件在屏幕上的位置。
- Value：显示滑动条的值，确定了可拖动滑块的位置。
- LeftValue：水平滑动条最左边的值。
- RightValue：水平滑动条最右边的值。
- Slider：水平滑动区域的样式，不使用则为默认样式。
- Thumb：用于显示可拖动滑块的 GUI 样式，不使用则为默认样式。

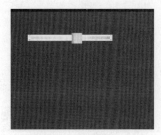

图 2-11　水平滑动条示例

下面制作一个水平滑动条，效果如图 2-11 所示。

源代码如下：

```
using UnityEngine;
using System.Collections;

public class UI_HorizontalSlider : MonoBehaviour {

    private float hSliderValue;
    private GUISkin skin;
    void Start ()
    {
        hSliderValue = 0.0F;
        skin = Resources.Load("Skin/MyGUISkin") as GUISkin;
    }

    void OnGUI ()
    {
        GUI.skin = skin;
        hSliderValue = GUI.HorizontalSlider(new Rect(25, 25, 100, 30), hSliderValue, 0.0F, 10.0F);

    }
}
```

在这个地方设置滑动条时不用 GUIStyle 这种方式，直接设置 Skin。设置过程如下：

首先在 Project 面板的 Resources/Skin 下建一个 GUI Skin 并命名为"MyGUISkin",如图 2-12、图 2-13 所示。

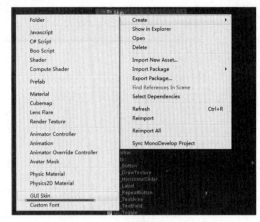

图 2-12　创建 GUISkin

图 2-13　GUISkin

设置水平滑动条选项,如图 2-14、图 2-15 所示。

垂直滑动条的开发与水平滑动条类似,使用到的函数如下:

GUI.VerticalSlider。

垂直滑动条风格设置页需要创建设置 GUI Skin。设置选项如图 2-16 所示。

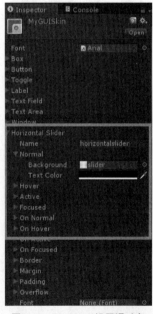

图 2-14　GUISkin 设置滑动条

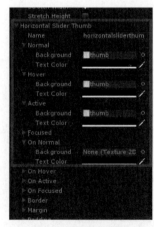

图 2-15　滑动条设置

图 2-16　滑动条设置

2.1.10　GUI.Window 窗口

常用的函数如下:

static function Window (id : int, clientRect : Rect, func : WindowFunction, text : String) : Rect
static function Window (id : int, clientRect : Rect, func : WindowFunction, text : String, style : GUIStyle) : Rect

参数说明：
- ID：每个窗口拥有唯一的 ID，注意在运行的时候不要有重复 ID 发生，否则会出现问题。
- ClientRect：窗口在屏幕上的位置以及大小。
- Func：可以在这个函数里面写一个 GUI 控件，从这里可以看出窗口实际上是一个容器。
- Text：窗口的名称。
- Style：窗口的样式。

下面制作一个窗口，效果如图 2-17 所示。
源代码如下：

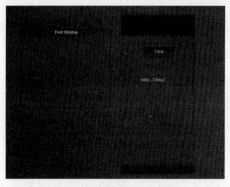

图 2-17　GUI.Window 示例

```
using UnityEngine;
using System.Collections;

public class UI_Window : MonoBehaviour
{
    // 窗体 1 的位置以及大小
    private Rect windowRect1;

    // 窗体 2 的位置以及大小
    private Rect windowRect2;
    // 窗体 2 的风格
    private GUIStyle window2Style;
    // 是否单击按钮
    private bool isClick = false;
    void Start ()
    {
        windowRect1 = new Rect(200, 100, 300, 250);
        windowRect2 = new Rect(530,50,256,512);
        window2Style = new GUIStyle();
        window2Style.normal.background = Resources.Load("Textures/Window/背景") as Texture2D;

    }



    void OnGUI ()
    {
        windowRect1 = GUI.Window(0, windowRect1, DoMyWindow1, "First Window");
        windowRect2 = GUI.Window(1, windowRect2, DoMyWindow2, "",window2Style);
    }

    void DoMyWindow1(int windowID)
    {


    }

    void DoMyWindow2(int windowID)
```

```
        {
            if(GUI.Button(new Rect(80,100,100,40),"Click"))
            {
                // 连续单击按钮可以切换 Label 的显示与隐藏
                isClick = !isClick;
            }

            if(isClick)
                GUI.Label(new Rect(70,200,120,40),"Hello , China !");

            // 拖拉窗口
            GUI.DragWindow();
        }
}
```

2.1.11 GUIContent.Tooltip 工具提示

这是非常实用的一个功能,当鼠标经过界面元素的时候会给出提示。下面制作一个按钮,当鼠标滑过按钮的时候给出一个提示,完成效果如图 2-18 所示。

图 2-18 工具提示示例

代码如下:

```
void OnGUI ()
{
    if(GUI.Button(new Rect(300, 300, 100, 20), new GUIContent("clik me", "这是一个按钮提示")))
    {

    }

    if(GUI.tooltip.Length>0)
        GUI.Label(new Rect(Input.mousePosition.x,Screen.height-Input.mousePosition.y, 200, 40), GUI.tooltip);

}
```

2.1.12 滚动视图

主要函数如下:

```
static function BeginScrollView (position : Rect, scrollPosition : Vector2, viewRect : Rect) : Vector2
static function EndScrollView () : void
```

参数说明:

第一个函数表示滚动视图开始,第二个函数表示滚动视图结束。其他需要滚动的内容放在两个函数之间。

- Position:滚动视图开始的位置以及大小。

- ScrollPosition：滚动的位置。
- ViewRect：滚动视图窗口内需要滚动的窗口的位置以及大小，当 ViewRect 的宽和高大于 Position 的宽度和高度的时候会显示滚动条。

下面制作一个滚动视图，效果如图 2-19 所示。

源代码如下：

```
using UnityEngine;
using System.Collections;

public class UI_ScrollView : MonoBehaviour {

    public Vector2 scrollPosition;
    // 导入图片
    private Texture2D tex;
    void Start ()
    {
        scrollPosition = Vector2.zero;
        tex = Resources.Load("Textures/ScrollView/花") as Texture2D;
    }

    void OnGUI ()
    {
        scrollPosition = GUI.BeginScrollView(new Rect(200, 100, 500, 450), scrollPosition, new Rect(0, 0, 960, 600));
        GUI.DrawTexture(new Rect(0,0,960,600),tex);
        GUI.EndScrollView();
    }
}
```

图 2-19　滚动视图示例

2.1.13　使用 Unity 自带的控件实现一个树形列表

树形列表非常有用，下面介绍一种实现树形列表的方法。

建立树形列表，OnGUI 实现的树形列表用一个多层循环来生成。

树形列表的效果图如图 2-20 所示。

源代码如下：

```
using UnityEngine;
using System.Collections;

public class UI_TreeList : MonoBehaviour
{

        // 状态数组，控制是否打开
        private bool[] state;
        // 滚动视图
        private   Vector2 scrollPosition;
        // 累计控件数目
```

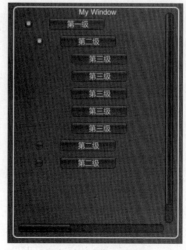

图 2-20　树形列表示例

```csharp
        private int number;
        public Rect windowRect = new Rect (20, 20, 300, 400);

        void Start ()
        {       // 默认不打开
                state = new bool[100];
                scrollPosition = Vector2.zero;
                number = 0;
        }

        /// <summary>
        /// 建立树形列表  /// </summary>
        public void buildTreeList (float x, float y)
        {
                float offset = 0;
                int index = 0;
                state [index] = GUI.Toggle (new Rect (x, y, 20, 20), state [index], "");
                if (GUI.Button (new Rect (x + 45, y, 100, 20), "第一级"))
                {
                }
                // 假如打开
                if (state [index])
                {
                        for (int i = 0; i<3; i++)
                        {
                                index++;
                                offset += 30;
                                state [index] = GUI.Toggle (new Rect (x + 20, y + offset, 20, 20), state [index], "");
                                if (GUI.Button (new Rect (x + 65, y + offset, 100, 20), "第二级"))
                                {
                                }
                                if (state [index])
                                {
                                        for (int ii = 0; ii<5; ii++)
                                        {
                                                index++;
                                                offset += 30;
                                                if (GUI.Button (new Rect (x + 85, y + offset, 100, 20), "第三级"))
                                                {
                                                }
                                        }
                                }
                        }
                }
                number = index;
        }

        void OnGUI ()
        {
                windowRect = GUI.Window (0, windowRect, DoMyWindow, "My Window");
```

```
        }
        void DoMyWindow (int windowID)
        {
            scrollPosition = GUI.BeginScrollView (new Rect (10, 10, 280, 380), scrollPosition, new Rect
(0, 0, 100 * number, 50 * number));
            buildTreeList (10, 10);
            GUI.EndScrollView ();

        }

}
```

2.1.14 基于 OnGUI 下的屏幕自适应

OnGUI 下的屏幕自适应是非常简单的。首先创建一个名为"GUIRoot.cs"的脚本，这个脚本里面默认以1366×768这个屏幕尺寸作为参考，超过或者小于该尺寸的屏幕会自动缩放。在1366×768这个尺寸下屏幕上显示的效果最好。这个参数可以自己定义。GUIRoot.cs 的脚本如下：

```
using UnityEngine;
using System.Collections;

public class GUIRoot : MonoBehaviour
{

    public static float wscale=1.0f;
    public static float hscale=1.0f;
    void Start ()
    {
        wscale=Screen.width/1366.0f;
        hscale=Screen.height/768.0f;




    }


    void Update ()
    {
        wscale=Screen.width/1024.0f;
        hscale=Screen.height/768.0f;




    }

}
```

创建完该脚本以后，需要在其他 OnGUI 函数里面加上一行代码，这行代码放在 OnGUI 函数首行，该行代码为：

```
GUI.matrix=Matrix4x4.TRS(Vector3.zero,Quaternion.identity,new Vector3(GUIRoot.wscale,GUIRoot.hscale,1));
```

示例请参考第二章工程文件 HelloUI 下的 Adaptation 场景。

2.1.15 制作一个简单的序列帧

创建一个 Cube，然后逐渐拉近与摄影机的距离，如图 2-21 所示。

创建一个名为"UI_Frame.cs"的脚本，编辑脚本如下：

```
using UnityEngine;
using System.Collections;

public class UI_Frame : MonoBehaviour {

    private int index = 0;
    private int delay = 0;
    private Texture2D[] frame;
    void Start ()
    {
        frame = new Texture2D[3];
        frame[0] = Resources.Load("Textures/序列帧/1") as Texture2D;
        frame[1] = Resources.Load("Textures/序列帧/2") as Texture2D;
        frame[2] = Resources.Load("Textures/序列帧/3") as Texture2D;
    }

    void OnGUI ()
    {
        delay++;
        // 每隔 45 帧切换图片
        if(delay%45 == 0)
        {
            index++;
            if(index>=frame.Length)
            {
                index = 0;
            }
        }
        GUI.DrawTexture(new Rect(0,0,446,279),frame[index]);
    }
}
```

图 2-21 序列帧要用的图片

将脚本拖到摄影机上运行，可以看到一个动态的 Cube 变化效果，如图 2-22 所示。

序列帧是很常用的功能，在有些时候序列帧动画可以代替视频，同时它也可以做出很多动态特效。

2.1.16 制作一个简单的动态柱状图

OnGUI 可以做出一些很有意思的界面效果，下面制作一个动态的柱状图。代码如下：

图 2-22 序列帧示例

```
using System.Collections;

public class BarGraph : MonoBehaviour
```

```csharp
{
    // 柱状图
    private float redBar_h = 0;
    private float blueBar_h = 0;
    private float yellow_h = 0;
    private float delay = 0;
    private bool isChange = false;
    private Texture2D redTex;
    private Texture2D blueTex;
    private Texture2D yellowTex;
    private Texture2D bg;
    void Start ()
    {
        redTex = Resources.Load("Textures/柱状图/red") as Texture2D;
        blueTex = Resources.Load("Textures/柱状图/blue") as Texture2D;
        yellowTex = Resources.Load("Textures/柱状图/yellow") as Texture2D;
        bg = Resources.Load("Textures/柱状图/bg") as Texture2D;
    }


    void OnGUI ()
    {
        if(GUI.Button(new Rect(10,10,100,30),"start"))
        {
            isChange = true;
        }
        if(isChange)
        {
            delay+=0.005f;
            if(delay>=1)
            {
                delay = 0;
            }
            Vector3 v = Vector3.Lerp(new Vector3(0,0,0),new Vector3(150,100,60),delay);
            redBar_h = v.x;
            blueBar_h = v.y;
            yellow_h = v.z;
        }

        GUI.DrawTexture(new Rect(10,61,353,206),bg);
        GUI.DrawTexture(new Rect(110,250,40,-redBar_h),redTex);
        GUI.DrawTexture(new Rect(170,250,40,-blueBar_h),blueTex);
        GUI.DrawTexture(new Rect(230,250,40,-yellow_h),yellowTex);


    }
}
```

这个脚本其实很简单，渐变效果主要使用 Vector3.Lerp 线性插值这个函数，注意这个函数是持续性的；在 GUI.DrawTexture (new Rect(110,250,40,-redBar_h),redTex);这段代码中会发现绘图图片的高度是负值，这表明增长方向从下往上和我们平时写的有点区别，单击"Start"，可以发现柱状条循环的由低变高。最后的效果如图 2-23 所示。

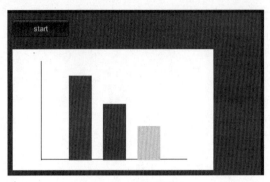

图 2-23　柱状图示例

2.1.17　制作一个图片查看器

下面介绍用 OnGUI 中的一些函数制作图片查看动态效果的方法，如图 2-24、图 2-25 所示。

图 2-24　图片查看器示例一

图 2-25　图片查看器示例二

创建一个名为"PhotoViewer.cs"的脚本，脚本片段 1 如下：

```csharp
using UnityEngine;
using System.Collections;

public class PhotoViewer : MonoBehaviour
{
    private Texture2D[] textures;
    private float hSbarValue = 0;
    private int texCount;
    private float delayOffset = 0;
    private float lastPos = 0;
    private bool isLeft = true;
    private Rect allTexShowWindowRect;
    void start ()
    {

        allTexShowWindowRect = new Rect (-10,50,1386,740) ;
        textures = new Texture2D[a] ;
        for(int 1=0;i<textures.Length;i++)
        {
            textures[i] = Resources.Load("Textures/图片查看器/"+(i+1)) as Texture2D;
        }

        texCount = textures.Length;

    }
```

脚本片段二如下：

```csharp
void OnGUI ()
{
    allTexShowWindowRect = new Rect(0,50,Screen.width,Screen.height-50);
    allTexShowWindowRect = GUI.Window(4, allTexShowWindowRect, DoAllTexWindow, "");

}
```

脚本片段三如下：

```csharp
void DoAllTexWindow(int windowID)
{

    if(textures!=null)
    {
        delayOffset+=0.01f;
        for(int i=0;i<textures.Length;i++)
        {
            if(textures[i]!=null)
            {

                int temp = (int)hSbarValue;
                if(temp == i)
                {
                    //600 400
```

```
                        // 长宽变化
                        Vector2 v = Vector2.Lerp(new Vector2(150,100),new
Vector2(600,400),delayOffset);
                        // 位置变化
                        Vector2 v1 = Vector2.Lerp(new Vector2(lastPos,500),new
Vector2(300,100),delayOffset);
                        GUI.DrawTexture(new Rect(v1.x,v1.y,v.x,v.y),textures[i]);

                    }
                    else
                    {
                        if(i<temp)
                        {
                            if(i == temp-1)
                            {
                                if(isLeft)
                                {
                                    // 位置变化
                                    Vector2 v1 = Vector2.Lerp(new Vector2(300,430),new
Vector2(140,430),delayOffset);
                                    GUI.DrawTexture(new Rect(v1.x,430,150,100),textures[i]);
                                }
                                else
                                {
                                    // delayOffset+=0.01f;
                                    // 长宽变化
                                    Vector2 v = Vector2.Lerp(new Vector2(600,400),new
Vector2(150,100),delayOffset);
                                    // 位置变化
                                    Vector2 v1 = Vector2.Lerp(new Vector2(300,100),new
Vector2(140,430),delayOffset);
                                    GUI.DrawTexture(new Rect(v1.x,v1.y,v.x,v.y),textures[i]);
                                }

                            }
                            else
                            {
                                Vector2 v2 = Vector2.Lerp(new Vector2(300-(temp-i-1)*160,200),new
Vector2(300-(temp-i)*160,430),delayOffset);
                                GUI.DrawTexture(new Rect(v2.x,430,150,100),textures[i]);

                            }
                        }
                        else if(i>temp)
                        {
```

```
                            if(i == temp+1)
                            {
                                if(isLeft)
                                {
                                    // 长宽变化
                                    Vector2 v = Vector2.Lerp(new Vector2(600,400),new Vector2(150,100),delayOffset);
                                    // 位置变化
                                    Vector2 v1 = Vector2.Lerp(new Vector2(300,200),new Vector2(910,430),delayOffset);
                                    GUI.DrawTexture(new Rect(v1.x,v1.y,v.x,v.y),textures[i]);
                                }
                                else
                                {
                                    // 位置变化
                                    Vector2 v1 = Vector2.Lerp(new Vector2(910+(i-temp),430),new Vector2(910,430),delayOffset);
                                    GUI.DrawTexture(new Rect(v1.x,430,150,100),textures[i]);
                                }
                            }
                            else
                            {
                                Vector2 v2 = Vector2.Lerp(new Vector2(910+(i-temp-1)*160,200),new Vector2(910+(i-temp-1)*160,430),delayOffset);
                                GUI.DrawTexture(new Rect(v2.x,430,150,100),textures[i]);
                            }

                        }
                    }
                }
            }
        }
    }
    if(GUI.Button(new Rect(Screen.width/2-100,400,100,30),"<-"))
    {
        hSbarValue -- ;
        if(hSbarValue<=0)
        {
            hSbarValue = 0;
```

```
            }
            delayOffset = 0;
            lastPos = 140;
            isLeft = true;
        }

        if(GUI.Button(new Rect(Screen.width/2+150,400,100,30),"->"))
        {
            hSbarValue++;
            if(hSbarValue>=texCount)
            {
                hSbarValue = texCount -1;
            }
            delayOffset = 0;

            lastPos = 750;
            isLeft = false;
        }
    }
}
```

2.1.18 制作一个小地图

在制作小地图时，先要建立适当的坐标系，然后确定游戏物体到起点的距离比，最后同比例绘制到界面上。

首先创建一个场景，这是一个使用 Terrain 创建的地形，在这个场景内放入一个 Box 表示 Player，如图 2-26 所示。

在 Scene 编辑面板下切换到俯视图，创建两个球体作为地图的起点和终点，如图 2-27 所示。

将起点和终点放在如上图位置是因为这样能使三维场景内的坐标系和界面的坐标系相似。

图 2-26　三维场景

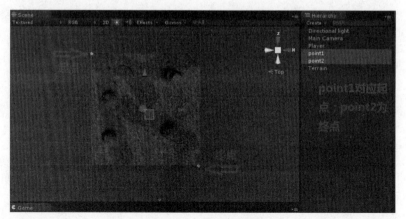

图 2-27　添加起点和终点

现在只要计算出 Player 到起点的横向距离与整个场景的宽度的比率以及 Player 到起点的纵向距离与整个场景的高度比率。当获得这个比率后就可以得到小地图上的 Player 标记点在小地图上距离起点的位置。最后大家注意一下，由于在界面上的小地图的起始点位置不一定是（0,0），所以在设置 Player 标记点的时候，位置要加上小地图的起始位置。

下面把 Player、point1、point2 这 3 个三维物体隐藏掉，然后调整到顶视图，将整个场景截图，如图 2-28 所示。同时制作一个红色的贴图放在工程中，如图 2-29 所示。

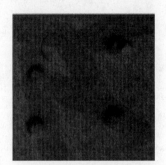

图 2-28　截取顶视图

图 2-29　制作小地图的图片资源文件

代码如下：

```csharp
using UnityEngine;
using System.Collections;

public class MiniMap : MonoBehaviour {
    // 三维场景中的起始点
    private Transform startPoint;
    // 三维场景结束点
    private Transform endPoint;
    // 动态目标物
    private Transform player;
    // 水平方向离起点的比率
    private float xScale;
    // 纵向离起点比率
    private float yScale;
    private float width;
    private float height;

    // 小地图贴图
    private Texture2D miniMapTex;
    // 红色标记点
    private Texture2D redPoint;

    // 小地图起始位置
    private float x =100;
    private float y =100;
    void Start ()
```

```
        {
                startPoint = GameObject.Find("point1") . transform;
                endPoint = GameObject.Find("point2") . transform;
                player = GameObject.Find("Player") . transform;

                width = Mathf.Abs(endPoint.position.x – startPoint.position.x);
                height = Mathf.Abs(endPoint.position.z – startPoint.position.z);

        miniMapTex = Resources.Load("Textures/小地图/minimap") as Texture2D;
        redPoint = Resources.Load("Textures/小地图/redPoint") as Texture2D;
}
void OnGUI ()
{
        // 横向比率
        xScale = Mathf.Abs(player.position.x – starPoint.position.x)/width;
        // 纵向比率
        yScale = Mathf.Abs(player.position.z – starPoint.position.z)/width;

        GUI.DrawTexture(new Rect(x,y,miniMapTex.width,miniMpTex.height),miniMapTex);
        GUI.DrawTexture(new Rect(x+miniMapTex.width*xScale,y+miniMapTex.height*yScale,10,10),redPoint);

}
```

2.2 NGUI

NGUI 是使用 Unity 开发 UI 时常用的一个插件，它拥有 Draw Call 低、可视化操作强的优点。

2.2.1 NGUI 概况

本书使用的是 NGUI Next-Gen UIv3.0.8 版本。

在将 NGUI-Next-Gen-UIv3.0.8.Unitypackage 资源包导入到工程文件夹中时，会在 Assets 下面多出一个 NGUI 的文件夹，如图 2-30 所示。

图 2-30　插件导入到 Unity 中

NGUI 里面的控件实际上并不是二维控件，而是三维物体，摄影机通过正交投射将控件渲染到屏幕上。NGUI 的摄影机的深度要大于三维场景中的摄影机深度，这是因为深度高的摄影机渲染的画面效果在深度低的摄影机渲染画面之上。这就是 NGUI 能够用三维物体创建出二维控件的原因。

2.2.2 NGUI 与 OnGUI 的差别

NGUI 摄影机的深度低于 OnGUI 摄影机深度。当 NGUI 与 OnGUI 混合使用的时候，特别要注意这一点。在同一位置的时候，OnGUI 的控件会遮挡 NGUI 控件。在 Draw Call 上，NGUI 要好一些，如果发布的平台对 Draw Call 要求很严格（如移动平台），那么 NGUI 可能更合适。如果觉得在三维场景里调整 NGUI 比较麻烦，那么可以使用 OnGUI 那种代码式的创建方式。总之，选择哪一个界面系统要根据开发的项目来看，二者并没有绝对的优劣。

第 3 章 多媒体应用

本章主要介绍音、视频加载。音、视频加载包括两个部分：本地加载和网络下载。视频方面除 Unity 提供的 MovieTexture 以外，还将介绍一个可以控制视频播放进度的插件。

3.1 音频的控制

3.1.1 本地音频加载与播放

Unity 支持.aif、.wav、.mp3、.ogg 等音频格式以及.xm、.mod、.it 和.s3m 等音频模块。最常用的有.wav、.mp3 和.ogg。

下载两个 mp3 文件，将它放到名为"MediaDemo"的工程文件中，如图 3-1 所示。

关于音频控制这里要用到两个类：Audio Source 和 Audio Clip，Audio Source 相当于一个唱片机，AudioClip 相当于唱片机里面的磁带。唱片机（Audio Source）用于控制磁带（Audio Clip）的播放、暂停、停止和替换等。如果想播放声音，就需要有 Audio Source 组件。如果声音必须要被听到，就还需要有监听声音的组件，即 Audio Listener。Audio Listener 一般默认存在于摄影机（Camera）上，在创建的摄影机中就能找到。

下面为场景中的摄影机添加 Audio Source 组件，关于声音播放控制，我们用代码实现，如图 3-2 所示为在摄影机上添加的 Audio Source。

图 3-1 MediaDemo

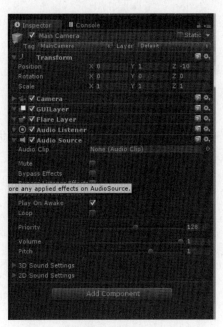

图 3-2 在摄影机上添加 Audio Source

创建一个名为"Audio Controller.cs"的脚本，将这个脚本添加到摄影机上，打开脚本添加如下代码：

```csharp
using UnityEngine;
using System.Collections;

public class AudioController : MonoBehaviour {

    private AudioClip audio1;
    private AudioClip audio2;

    private AudioSource source;

    private bool isPlayAudio1;
    private bool isPlayAudio2;

    private string audio1Str;
    private string audio2Str;

    private string hintStr;
    void Start ()
    {
        audio1 = Resources.Load("Audio/audio1") as AudioClip;
        audio2 = Resources.Load("Audio/audio2") as AudioClip;

        source = GameObject.Find("Main Camera").GetComponent<AudioSource>();

        isPlayAudio1 = false;
        isPlayAudio2 = false;

        audio1Str = "播放 audio1";
        audio2Str = "播放 audio2";
        hintStr = "";
    }


    void OnGUI ()
    {
        if(GUI.Button(new Rect(300,100,200,60),audio1Str))
        {
            source.clip = audio1;
            audio2Str = "播放 audio2";
            isPlayAudio1 = !isPlayAudio1;
            if(isPlayAudio1)
            {
                audio1Str = "暂停 audio1";
                source.Play();
                hintStr = "audio1 正在播放。。。 ";
            }
            else
            {
                audio1Str = "播放 audio1";
                source.Pause();
                hintStr = "audio1 已经暂停。。。 ";
```

```
            }
        }
        if(GUI.Button(new Rect(550,100,200,60),audio2Str))
        {
            source.clip = audio2;
            audio1Str = "播放 audio1";
            isPlayAudio2 = !isPlayAudio2;
            if(isPlayAudio2)
            {
                audio2Str = "暂停 audio2";
                source.Play();
                hintStr = "audio2 正在播放。。。 ";
            }
            else
            {
                audio2Str = "播放 audio2";
                source.Pause();
                hintStr = "audio2 已经暂停。。。 ";
            }
        }
        GUI.Label(new Rect(350,200,200,60),hintStr);
    }
}
```

完成效果如图 3-3 所示。

除了上面这种动态加载本地音频的方法外，还可以用第 1 章介绍的方法直接将一个音频拖放到 Audio Source 下面的 Audio Clip 选项中，如图 3-4 所示。

图 3-3　切换声音

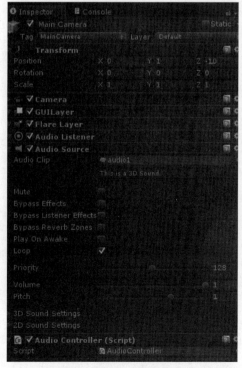

图 3-4　将声音直接拖到面板上

3.1.2 通过网络加载音频

如果要加载来自某个服务器下面的音频资源，就需要用到 Unity 中的 WWW 类。WWW 可以自动下载来自给定 URL 的音频资源。其音频格式必须为.ogg。下面用 WWW 加载声音，代码如下：

```csharp
using UnityEngine;
using System.Collections;

public class AudioFromWWW : MonoBehaviour {

    private AudioClip audio;

    private AudioSource source;

    private bool isPlayAudio;

    private string audioStr;

    IEnumerator Start ()
    {
        source = GameObject.Find("Main Camera").GetComponent<AudioSource>();
        isPlayAudio = false;
        audioStr = "播放";

        string url = "file:// "+Application.dataPath+"/Data/Audio/audio1.ogg";

        yield return StartCoroutine(LoadAudio(url));
    }

    void OnGUI ()
    {
        if(GUI.Button(new Rect(300,200,200,70),audioStr))
        {
            if(source.clip!=null)
                source.Play();
        }
    }

    IEnumerator LoadAudio(string url)
    {
        WWW www = new WWW(url);
        yield return www;
        source.clip = www.audioClip;

    }
}
```

上面这段代码是下载完毕以后再播放,当然还可以边下载边播放,这种方式可以参考第 4 章数据加载与卸载中的内容。源代码对应第 2 章 MediaDemo 工程文件夹下的 AudioWWW 场景,效果如图 3-5 所示。

图 3-5　WWW 来加载声音文件

3.2　视频播放控制

3.2.1　MovieTexture 的视频播放控制

　　MovieTexture 是 Unity 中提供的用于视频播放控制的类,它继承于 Texture,所以既可以用于 UI 做二维视频播放,也可以用于游戏物体的贴图上。MovieTexture 不支持 Android 和 IOS,可以使用 iPhoneUtils.PlayMovie 和 iPhoneUtils.PlayMovieURL 来控制播放视频。

　　MovieTexture 支持的视频格式有:.mov、.mpg、.mpeg、.mp4、.avi、.asf、.ogg。如果使用 WWW 加载视频,那么视频格式就只能是.ogg。

　　1. 在二维界面上使用 MovieTexture

代码如下:

```
using System.Collections;

public class MovieTextureDemo : MonoBehaviour {

    public MovieTexture movie;
    private string str = "暂停视频";
    private bool isPlay = true;
    void Start ()
    {
        movie.Play();
        movie.loop = true;
    }


    void OnGUI ()
    {
        GUI.DrawTexture(new Rect(300,100,500,350),movie);

        if(GUI.Button(new Rect(500,500,100,60),str))
```

```
            {
                isPlay = !isPlay;
                if(isPlay)
                {
                    movie.Play();
                    str = "暂停视频";
                }
                else
                {
                    movie.Pause();
                    str = "打开视频";
                }
            }
        }
}
```

将这段代码拖到 Main Camera 上作为它的一个组件，并为 Main Camera 添加 Audio Source 组件。将视频的声音添加给 Audio Source 的 Audio Clip，设置界面如图 3-6 所示。对应的资源目录如图 3-7 所示。

图 3-6　视频播放

图 3-7　资源目录

2. 将 MovieTexture 作为一个三维物体的贴图

在场景中建立一个面片，然后新建一个带有视频贴图的材质，将这个材质赋予面片。控制视频播放的方式有播放、暂停、停止，分别对应 Play()、Pause()和 Stop()这三个函数。MovieTexture 还有一些其他属性和函数，可以查阅 Unity 的脚本帮助文件。

如图 3-8 所示就是一个带视频贴图的三维面片。

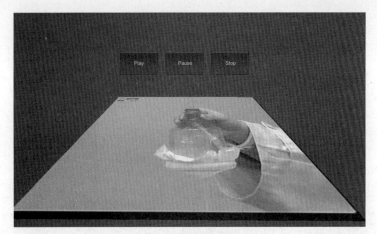

图 3-8 控制播放视频

代码如下：

```
using UnityEngine;
using System.Collections;

public class MovieFor3D : MonoBehaviour {

    private GameObject movieObj;
    private MovieTexture movieTexture;
    private AudioSource audioSource;
    void Start ()
    {
        movieObj = GameObject.Find("Plane");
        // 获取 Plane 材质上面的贴图
        movieTexture = (MovieTexture)movieObj.renderer.material.mainTexture;

        // 获取已经添加到摄影机上的声音播放器
        audioSource = GameObject.Find("Main Camera").audio;
        audioSource.clip = movieTexture.audioClip;
    }

    // Update is called once per frame
    void OnGUI ()
    {
        if(GUI.Button(new Rect(360,150,100,60),"Play"))
        {
            // 播放视频和声音
            movieTexture.Play();
            audioSource.Play();
        }

        if(GUI.Button(new Rect(480,150,100,60),"Pause"))
        {
            // 暂停视频和声音
            movieTexture.Pause();
            audioSource.Pause();
```

```
            }
            if(GUI.Button(new Rect(600,150,100,60),"Stop"))
            {
                // 停止视频和声音
                movieTexture.Stop();
                audioSource.Stop();
            }
        }
    }
```

设置界面如图 3-9 所示。

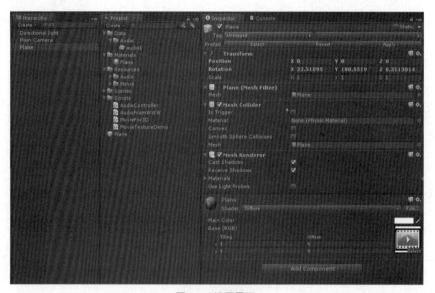

图 3-9 设置界面

3.2.2 AvPro QuickTime 的视频播放

AvPro QuickTime 能够调节播放进度,它支持本地和网络视频播放。它支持的格式比较多,推荐使用 mp4 和 MOV。AvPro QuickTime 需要的资源要多一些,有时候会让帧率变得很低。

在使用 AvPro QuickTime 插件前需要安装 QuickTime 播放器,如图 3-10、图 3-11 所示为安装截图。

图 3-10 安装播放器(1)

图 3-11 安装播放器(2)

完成后可以导入 AvPro QuickTime 插件包，如图 3-12 所示。打开场景，按 Play 运行，如图 3-13 所示。

图 3-12　导入插件包

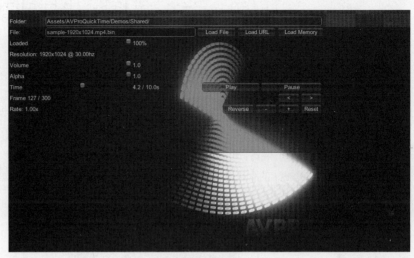

图 3-13　运行

第 4 章　数据加载与卸载

本章主要介绍使用 Resource 和 WWW 加载资源的方法及 Unity 的内存管理。

4.1　Resource.Load 加载资源

用 Resources.Load 函数访问的资源，需要把资源放在 Assets 中名为"Resources"的文件夹下，它能加载返回值为 Object 的资源，如图片、视频、声音、游戏物体等，如图 4-1 所示。

图 4-1　Resources 下的资源

下面是导入图片、声音和视频的一个案例，主要代码如下：

```
// 图片
private Texture2D tex;
// 声音
private AudioClip clip;
// 视频
private MovieTexture movie;

private bool isShowMovie = false;
private bool isPlayAudio = false;
void Start ()
{
    // 导入图片
    tex = Resources.Load("Images/Flower") as Texture2D;
    // 导入声音
    clip = Resources.Load("Audio/audio1") as AudioClip;
    // 导入视频
    movie = Resources.Load("Movie/movie") as MovieTexture;
```

```
        movie.Play();
        movie.loop = true;
        // 这样调用是因为脚本所在的这个游戏物体有 AudioSource 这个组件
        gameObject.audio.clip = movie.audioClip;
}


void OnGUI ()
{
        GUI.DrawTexture(new Rect(10,10,200,200),tex);
        if(isShowMovie)
                GUI.DrawTexture(new Rect(230,10,200,200),movie);
        if(GUI.Button(new Rect(250,230,200,60),"click"))
        {
                isShowMovie = !isShowMovie;
                if(isShowMovie)
                {
                        gameObject.audio.clip = movie.audioClip;
                        gameObject.audio.Play();
                }
                else
                {
                        gameObject.audio.Stop();
                }
        }
        if(GUI.Button(new Rect(570,230,200,60),"play auido"))
        {

                gameObject.audio.clip = clip;
                gameObject.audio.Play();
        }
}
```

效果如图 4-2 所示。

图 4-2　Resources.Load 导入图片、声音、视频

4.2 WWW 加载

使用 WWW 加载资源时，WWW(url)会在后台开始下载，并且返回一个新的 WWW 对象。这里主要介绍用同步和异步加载视频和模型。

1. 同步加载视频

代码如下：

```
private MovieTexture movieTexture;
private string url;
void Start ()
{
    if(Application.isWebPlayer)
    {
        // 要使用绝对路径
        url = Application.dataPath+"/Data/Movie/movie.ogg";
        Debug.Log(url);
    }
    else
    {
        // www 在编辑器和发布为单机版注意路径
        url = "file://"+Application.dataPath+"/Data/Movie/movie.ogg";

    }

    StartCoroutine(SynchronousLoadMovie(url));
}


void OnGUI ()
{
    if(movieTexture!=null)
    {

    GUI.DrawTexture(new Rect(200,100,500,450),movieTexture);
    }

}

// 同步加载视频
IEnumerator SynchronousLoadMovie(string url)
{
    movieTexture = null;
    yield return Resources.UnloadUnusedAssets();
    // 释放掉没用的内存空间
    WWW www = new WWW(url);
    yield return www;
    movieTexture = www.movie;
    www = null;
    movieTexture.Play();
```

第 **4** 章 数据加载与卸载

```
    }
}
```

2. 异步加载视频

代码如下：

```
IEnumerator AsyncLoadMovie(string url)
{
    yield return Resources.UnloadUnusedAssets();
    WWW www = new WWW(url);
    movieTexture = www.movie;
    while (!movieTexture.isReadyToPlay)
        yield return 0;
    gameObject.audio.clip = movieTexture.audioClip;

    // Play both movie & sound
    // 播放音频和声音
    movieTexture.Play();
    gameObject.audio.Play();

}
```

3. 同步加载模型

在 Data/Model 下有一个名为"Cube"的 Prefab 的文件，如图 4-3 所示。这里使用打包插件 ExportAssetBundles.cs，注意要把这个文件放到 Editor 下，如图 4-4 所示。

添加完成以后，可以在 Unity 的菜单栏 Assets 中看到如图 4-5 所示的几个选项。

图 4-3 加载模型

图 4-4 打包插件

单击选中 Cube，然后选择第一种方式打包 Cube.prefab，如图 4-6 所示。

67

图 4-5 打包选项

图 4-6 将 Cube 打包

打包完成后会有两个文件，其中后缀名为.unity3d 的文件就是打包后的资源文件。下面将资源文件加载到场景中去。

代码如下：

```
// 导入模型资源
IEnumerator LoadAsset(string url)
{
        WWW www = new WWW (url);
    // 等待下载完成
    yield return www;
    // 获取指定的主资源并实例化
    AssetBundle asset = www.assetBundle;
    //
    GameObject obj = Instantiate(asset.Load("Cube")) as GameObject;
    // 给复制的物体重命名
    obj.name = obj.name.Replace("(Clone)","");
    // 确保内存中只有一个复制
    asset.Unload(false);//卸载 Assetbundle
    www = null;
    Resources.UnloadUnusedAssets();

}
```

Resource 加载方式必须要从一个固定的文件夹导入；WWW 可以从任意一个路径导入，无论是从本地还是从网络导入，这种区别也就导致 Resource 导入的资源要和程序文件组合在一起，而 WWW 导入的资源可以和主程序分离。采取何种方式导入资源，取决于项目需要。

第 5 章　Unity 读写外部数据

本章主要介绍对一个 Xml 文件进行读写操作以及通过 Unity 对数据库进行增、删、改、查等操作的方法。

5.1　操作 Xml

5.1.1　C#操作 Xml 文件基础知识

已知有一个 Xml，文件名为"UserInfo.xml"，操作如下：

```
<?xml version="1.0" encoding="UTF-8" standalone="no"?>
<users>
    <user1 姓名="张三" 密码="123">我是张三</user1>
    <user2 姓名="李四" 密码="123">我是李四</user2>
</users>
```

1. 插入一个节点，添加属性，保存(注意 xml 文件存放的路径)

```
XmlDocument xmlDoc=new XmlDocument();
xmlDoc.Load("c:/UserInfo.xml");
XmlNode root=xmlDoc.SelectSingleNode("users");// 查找
XmlElement xe1=xmlDoc.CreateElement("user3");// 创建一个节点
xe1.SetAttribute("姓名", "王二");// 设置该节点姓名属性
xe1.SetAttribute("密码", "123");// 设置该节点密码属性
xe1.InnerText="我是王二";
root.AppendChild(xe1);// 添加到节点中
xmlDoc.Save("c:/UserInfo.xml");
```

运行完成效果：

```
<?xml version="1.0" encoding="UTF-8" standalone="no"?>
<users>
    <user1 姓名="张三" 密码="123">我是张三</user1>
    <user2 姓名="李四" 密码="123">我是李四</user2>
    <user3 姓名="王二" 密码="123">我是王二</user3>
</users>
```

2. 修改节点

将 user3 密码修改为"1"内容修改为"我是王小二"
```
XmlDocument xmlDoc=new XmlDocument();
xmlDoc.Load("c:/UserInfo.xml");
XmlNode root=xmlDoc.SelectSingleNode("users");//查找
XmlElement user3 = (XmlElement)root.SelectSingleNode("user3");
user3.SetAttribute("密码","1");
```

```
user3.InnerText = "我是王二小";
xmlDoc.Save("c:/UserInfo.xml");
```

运行完成效果：

```xml
<?xml version="1.0" encoding="UTF-8" standalone="no"?>
<users>
  <user1 姓名="张三" 密码="123">我是张三</user1>
  <user2 姓名="李四" 密码="123">我是李四</user2>
  <user3 姓名="王二" 密码="1">我是王二小</user3>
</users>
```

3. 删除属性和节点

```csharp
XmlDocument xmlDoc=new XmlDocument();
xmlDoc.Load("c:/UserInfo.xml");
XmlNode root=xmlDoc.SelectSingleNode("users");// 查找
// 移除 user3 的密码这个属性
XmlElement user3 = (XmlElement)root.SelectSingleNode("user3");
        user3.RemoveAttribute("密码");
        // 移除 user2 这个节点
        root.RemoveChild(root.SelectSingleNode("user2"));
        xmlDoc.Save("c:/UserInfo.xml");
```

效果如下：

```xml
<?xml version="1.0" encoding="UTF-8" standalone="no"?>
<users>
  <user1 姓名="张三" 密码="123">我是张三</user1>
  <user3 姓名="王二">我是王二小</user3>
</users>
```

4. 遍历节点

```csharp
XmlDocument xmlDoc=new XmlDocument();
xmlDoc.Load("c:/UserInfo.xml");
XmlNode root=xmlDoc.SelectSingleNode("users");// 查找
// 遍历节点
foreach(XmlNode xn in root.ChildNodes)
    Debug.Log(xn.InnerText);
```

效果如图 5-1 所示。

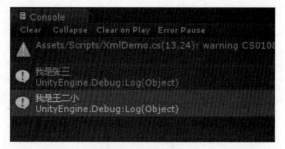

图 5-1　显示结果

5.1.2　Unity 加载 Xml 文件的方式

1. Unity 加载本地 Xml 文件

xmlDocument.Load("文件绝对路径");

2. Unity 加载异地 Xml 文件

在本地计算机上安装一个 Xampp，该软件里面集成了 Apache 服务器，将一个 Xml 文件放到服务器下（Apache 的文件放在 Htdocs 文件夹下），如图 5-2 所示。

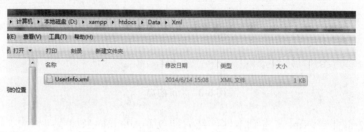

图 5-2　Xml 文件

Xml 内容如下：

```
void isUserExist(string name,string pwd)
{
    XmlElement root = readXmlFile.GetRootNode();

    for(int i=0;i<root.ChildNodes.Count;i++ )
    {
        XmlElement xe = (XmlElement)root.ChildNodes[i];

        if(name.Equals(xe.GetAttribute("姓名")))
        {
            user = xe;
        }
    }

    if(user != null)
    {
        if(pwd.Equals(user.GetAttribute("密码")))
        {
            // 提示信息
            loginStr = "登录成功";
            // 登录成功
            isLoginSucess = true;
        }
        else
        {
            loginStr = "密码错误！";
            // 登录失败
            isLoginSucess = false;
        }
    }
}
```

```csharp
            else
            {
                // 登录失败
                isLoginSucess = false;
                loginStr = "没有该用户！";
            }
        }
```

效果如图 5-3 所示。

```
1 <?xml version="1.0" encoding="UTF-8" standalone="no"?>
2 <users>
3       <user1 姓名="张三" 密码="123">我是张三</user1>
4       <user2 姓名="李四" 密码="123">我是李四</user2>
5 </users>
```

图 5-3 Xml 内容

先用 WWW 方式下载 Xml 文本到本地，然后通过 LoadXml()来加载，代码如下：

```csharp
// 注意一下路径，不包括 hodocs，网址写你服务器网址
private string url = "http://127.0.0.1/Data/Xml/UserInfo.xml";
    private string xmlText = "";
    private ReadXmlFile readXmlFile;
    IEnumerator Start()
    {
// 先将 Xml 下载下来
        WWW www = new WWW(url);
        yield return www;
        xmlText  = www.text;
        readXmlFile = new ReadXmlFile();
// 导入 Xml 文本
        readXmlFile.LoadXml(xmlText);

// 打印根节点下子节点的姓名这个属性
        foreach(XmlNode xn in readXmlFile.GetRootNode().ChildNodes)
        {
            Debug.Log(((XmlElement)xn).GetAttribute("姓名"));
        }

    }
```

效果如图 5-4 所示。

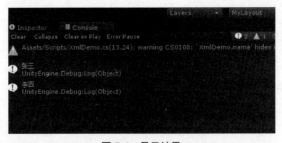

图 5-4 显示结果

5.1.3 Unity 与 Xml 交互案例：用户登录验证

下面是一个名为 UserInfo.xml 的内容：

```xml
<?xml version="1.0" encoding="UTF-8" standalone="no"?>
<users>
    <user1 姓名="张三" 密码="123">我是张三</user1>
    <user2 姓名="李四" 密码="123">我是李四</user2>
</users>
```

上面的 Xml 有 2 层，父节点叫作 users，子节点有两个，分别为 user1 和 user2，每个子节点有两个属性：姓名和密码。每个节点有一段文字内容。要求能够读取每个节点的名称内容和属性，并且能够添加节点和属性。

下面是一个操作 Xml 读取的类，默认继承 Object，这个类里面主要有 3 个函数：

- Load（string filePath）：通过文件路径来加载 Xml 文件内容。
- LoadXml（string xmlFile）：通过加载 Xml 文件内的文本内容加载 Xml。
- GetRootNode()：用来获取 Xml 的根节点，有了这个根节点，就可以访问每一个子节点。

代码如下：

```csharp
using System;
using System.IO;
using System.Xml;// 注意要添加这个命名空间
public class ReadXmlFile
{
    XmlDocument xmlDocument;
    public ReadXmlFile()
    {
        xmlDocument = new XmlDocument();
    }

    /// <summary>
    /// 通过文件路径导入 Xml 文件，用于单机版
    /// </summary>
    public bool Load(string filePath)
    {
        xmlDocument.Load(filePath);
        return true;
    }

    /// <summary>
    /// 通过加载 XML 文本内容加载 Xml，用于网页版
    /// </summary>
    public bool LoadXml(string xmlFile)
    {
        xmlDocument.LoadXml(xmlFile);
        return true;
    }

    // 获取根节点
```

```
public XmlElement GetRootNode()
    {
        return xmlDocument.DocumentElement;
    }
}
```

下面介绍一个登录验证的范例。

先建立一个 UI 界面，如图 5-5 所示。

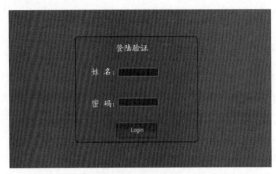

图 5-5　登录验证

部分主要代码如下：

```
void OnGUI ()
    {
        GUI.matrix=Matrix4x4.TRS(Vector3.zero,Quaternion.identity,new Vector3(GUIRoot.wscale,GUIRoot.hscale,1));
        windowRect = GUI.Window(0, windowRect, DoMyWindow, "");

    }


    void DoMyWindow(int windowID)
    {
        GUI.Label(new Rect(100,5,100,50),"登录验证",labelStyle);

        GUI.Label(new Rect(40,65,60,40),"姓　名：",labelStyle);
        name = GUI.TextField (new Rect(105,75, 100, 20), name, 15);

        GUI.Label(new Rect(40,130,60,60),"密　码：",labelStyle);
        pwd = GUI.TextField (new Rect(105,145, 100, 20), pwd, 15);

        if(GUI.Button(new Rect(100,200,100,40),"Login"))
        {

        }
    }
```

将 Xml 文件加载进来的代码如下：

```
readXmlFile = new ReadXmlFile();
// 发布为 EXE 访问本地 Xml 的时候使用这种加载方式,这种方式发布为网页不支持
readXmlFile.Load(Application.dataPath+"/Data/Xml/UserInfo.xml");
```

开始登录验证,首先判断是否存在这个用户,如果存在获取这个 Xml 节点,将节点密码这个属性与用户输入的属性进行比较,相同则登录成功,否则失败。

完整的 XmlDemo.cs 代码如下:

```csharp
using UnityEngine;
using System.Collections;
using System.Xml;
public class XmlDemo : MonoBehaviour {

    private ReadXmlFile readXmlFile;

    private Rect windowRect = new Rect(350, 200, 300, 300);
    private Font kaiTi;
    private GUIStyle labelStyle;


    private string name = "";
    private string pwd = "";

    // 判断是否登录成功
    private bool isLoginSucess = false;
    // 用户节点,如果为空说明用户不存在
    private XmlElement user;
    // 登录信息提示
    private string loginStr = "";
    void Start ()
    {

        kaiTi = Resources.Load("Font/simkai") as Font;
        labelStyle = new GUIStyle();
        labelStyle.font = kaiTi;
        labelStyle.alignment = TextAnchor.MiddleLeft;
        labelStyle.fontSize = 20;
        labelStyle.normal.textColor = Color.white;

        readXmlFile = new ReadXmlFile();
        // 发布为 EXE 访问本地 XML 的时候使用这种加载方式,这种方式发布为网页不支持
        readXmlFile.Load(Application.dataPath+"/Data/Xml/UserInfo.xml");

        user = null;



    }


    void OnGUI ()
    {
```

```
                GUI.matrix=Matrix4x4.TRS(Vector3.zero,Quaternion.identity,new
Vector3(GUIRoot.wscale,GUIRoot.hscale,1));
                windowRect = GUI.Window(0, windowRect, DoMyWindow, "");

        }


        void DoMyWindow(int windowID)
        {
            if(!isLoginSucess)
            {
                GUI.Label(new Rect(100,5,100,50),"登录验证",labelStyle);

                GUI.Label(new Rect(40,65,60,40),"姓 名：",labelStyle);
                name = GUI.TextField (new Rect(105,75, 100, 20), name, 15);

                GUI.Label(new Rect(40,130,60,60),"密 码：",labelStyle);
                pwd = GUI.TextField (new Rect(105,145, 100, 20), pwd, 15);

                if(GUI.Button(new Rect(100,200,100,40),"Login"))
                {
                    isUserExist(name,pwd);
                }

                GUI.Label(new Rect(80,255,130,40),loginStr,labelStyle);
            }
            else
            {
                GUI.Label(new Rect(80,100,130,40),user.InnerText,labelStyle);
                GUI.Label(new Rect(80,170,130,40),loginStr,labelStyle);
            }



        }


        void isUserExist(string name,string pwd)
        {
            XmlElement root = readXmlFile.GetRootNode();

            for(int i=0;i<root.ChildNodes.Count;i++ )
            {
                XmlElement xe = (XmlElement)root.ChildNodes[i];

                if(name.Equals(xe.GetAttribute("姓名")))
                {
                    user = xe;
                }
            }

            if(user != null)
            {
```

```
                if(pwd.Equals(user.GetAttribute("密码")))
                {
                    // 提示信息
                    loginStr = "登录成功";
                    // 登录成功
                    isLoginSucess = true;
                }
                else
                {
                    loginStr = "密码错误！";
                    // 登录失败
                    isLoginSucess = false;
                }
            }
            else
            {
                // 登录失败
                isLoginSucess = false;
                loginStr = "没有该用户！";
            }

        }
```

5.2 操作数据库

5.2.1 Xampp 介绍以及安装

Xampp（Apache+MySQL+PHP+PERL）是一个功能强大的建站集成软件包。通过 Xampp 集成的 Apache 和 MySQL 可以实现 Unity 与数据库的交互。

下面介绍 Xampp 的安装以及设置方法。

从网上下载 Xampp，单击安装，如图 5-6 所示。单击"OK"，选择储存路径后，单击"Next"，如图 5-7 所示。

图 5-6　安装 xampp

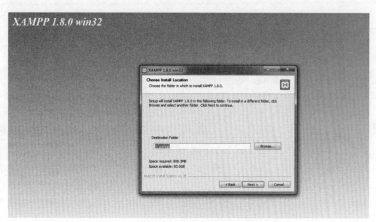

图 5-7　安装 xampp

可以将 Xampp 上的所有服务都装上，这里只装了 Apache 和 MySQL，单击"Install"，如图 5-8 所示。单击"Finish"安装完成，如图 5-9 所示。完成以后打开控制面板，如图 5-10、图 5-11 所示。

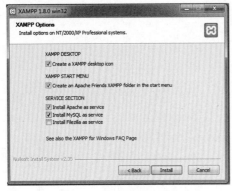

图 5-8　安装 xampp（1）

图 5-9　安装 xampp（2）

图 5-10　安装 xampp（3）

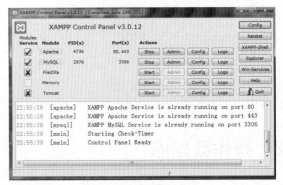

图 5-11　安装 xampp（4）

单击"Admin"可以打开一个网页，这个网页用于管理服务器和数据库。数据库和服务器默认密码为空，可以为数据库和服务器设置密码，也可以通过浏览器输入 http://localhost 进入 Xampp 管理页面，如图 5-12 所示。

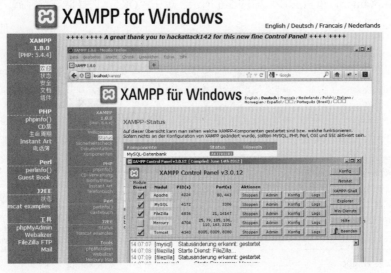

图 5-12　打开页面

单击左边的"安全"选项也可以设置密码，这里设置 MySQL 的用户名为"root"，密码为 qwer1234，如图 5-13 所示。

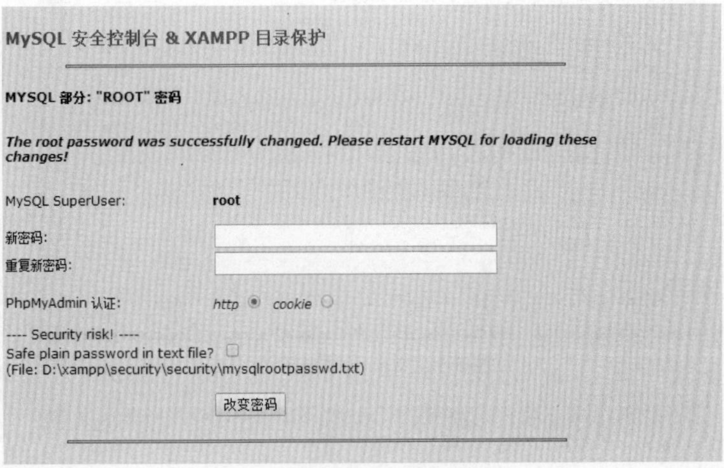

图 5-13 设置用户名和密码

设置完以后，在控制面板中要重启一下数据库和服务器（如果都设置了的话）。如果要放一些资源，如视频、文档、图片等在服务器下，可以将其放在"\xampp\htdocs"下，同样要重启一下服务器。

5.2.2 在 Xampp 上建立一个数据库

打开浏览器，在地址栏输入 http://localhost 进入 Xampp 首页，如图 5-14 所示。单击 phpMyAdmin 就会跳出一个窗口，输入用户名和密码后就可以进入 MySQL（如果设置了用户名和密码）。

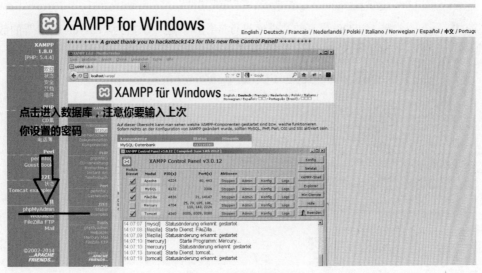

图 5-14 设置数据库密码

也可以通过 Xampp 的控制面板进入 MySQL，只要单击 MySql 一行的"Admin"，如图 5-15 所示。进入数据库后，可以设置文字字体等，如图 5-16 所示。

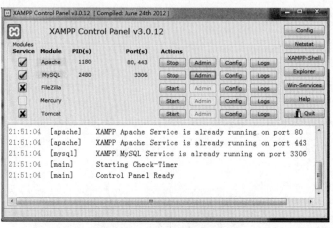

图 5-15　Xampp 控制面板

图 5-16　数据库页面

设置完以后，不要对 MySQL 已有的数据库进行操作，新建一个名为"easydatabase"的数据库（可以直接单击 SQL 用 SQL 语句新建），如图 5-17 所示。

图 5-17　在页面上创建新数据库

完成新建以后，单击 创建一个名为"学生信息"的表，并且输入 3 条信息。
SQL 语句如下：

```
CREATE TABLE 学生信息
(
    学号  varchar(14)    PRIMARY KEY,
    姓名  varchar(8) UNIQUE NOT NULL
);

INSERT INTO 学生信息 (学号, 姓名) VALUES
(001, "张三"),
(002, "李四"),
(003, "王五");
```

在网页中就会显示如图 5-18 所示的结果。

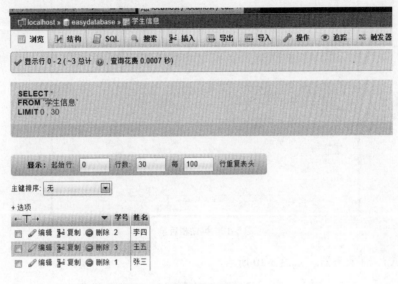

图 5-18 结果

5.2.3 创建一个 PHP 文件连接数据库

PHP 是一种创建动态交互性站点的强有力的服务器端脚本语言。

下面是连接 easydatabase 的 PHP 代码：

```
<?php
    $db=mysql_connect('localhost','root','qwer1234') or die('Unable to connect:'.mysql_error());
    mysql_query("SET NAMES 'utf8'");
    $b = mysql_select_db('easydatabase',$db) or die('Unable to connect to the database');
    $sql = $_POST['sql'];
    $result = mysql_query($sql) or die('Query failed!'.mysql_error());
    $info[]=array();
    $i=0;
    while($rs = mysql_fetch_assoc($result))
    {
        $info[$i]=$rs;
        $i++;
    }
    $infostr=json_encode($info);
```

```
$infostr= preg_replace("#\\\u([0-9a-f]{4})#ie", "iconv('UCS-2BE', 'UTF-8', pack('H4', '\\1'))", $infostr);
echo $infostr;
mysql_close($db);
?>
```

将这个 PHP 脚本命名为"connect.php",然后把这个脚本放到 Apache 服务器下(目录为 Xampp 下的 htdocs),如图 5-19 所示。

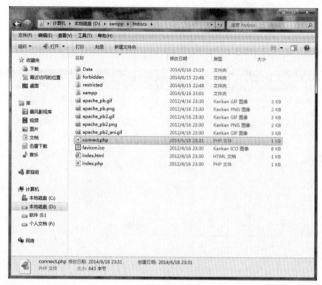

图 5-19 网站根目录

然后重启 Apache 服务器,如图 5-20 所示。

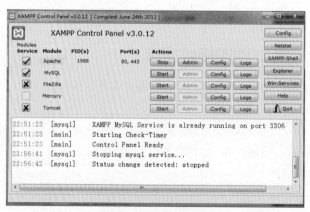

图 5-20 重启服务器

5.2.4 Unity+PHP+MySQL 操作数据库

想要操作数据库,需要先了解 Unity 提供的一些类和函数。下面主要介绍 WWW 和 WWWForm 网页表单。

1. WWW

WWW 是一个访问网页的类,可以通过 WWW(url)在后台开始下载,并且返回一个新的 WWW

对象，WWW 类可以用来发送 GET 和 POST 请求到服务器，WWW 类默认使用 GET 方法，如果提供一个 postData 参数还可以用 POST 方法。

2. WWWForm

WWWForm 用来生成表单数据，它是 postData 参数，可以构建表单数据。使用 WWW 类传递到 Web 服务器。

下面介绍用 WWW 访问服务器上的 connect.php 页面，然后用 WWWForm 提交需求的方法。

通过页面发送需求的代码如下：

```
public static string conUrl = "http://localhost/connect.php";
// 从数据库获取的结果
public static string result;
void Start ()
{
    result = "";
}
public static IEnumerator LoadData(string sql)
{

        WWWForm form = new WWWForm();
        // 提交 SQL 语句
        form.AddField("sql",sql);
        // 向服务器提交需求
        WWW hspost = new WWW(conUrl,form);
        yield return hspost;
        if(hspost.error != null)
        {
            print(hspost.error);
        }
        else
        {
            result = hspost.text;

        }
}
```

新建 DatabaseDemo.cs，添加一个名为 SelectAllInfo（）的函数，这个函数获取 easydatabase 中学生信息表中的所有信息，代码如下：

```
private string infoStr = "";
void Start ()
{
    StartCoroutine(SelectAllInfo());
}

IEnumerator SelectAllInfo()
{
    yield return StartCoroutine(OperateDatabase.LoadData("select * from 学生信息"));
    infoStr = OperateDatabase.result;
    Debug.Log(infoStr);
}
```

查看打印出来的信息，如图 5-21 所示。

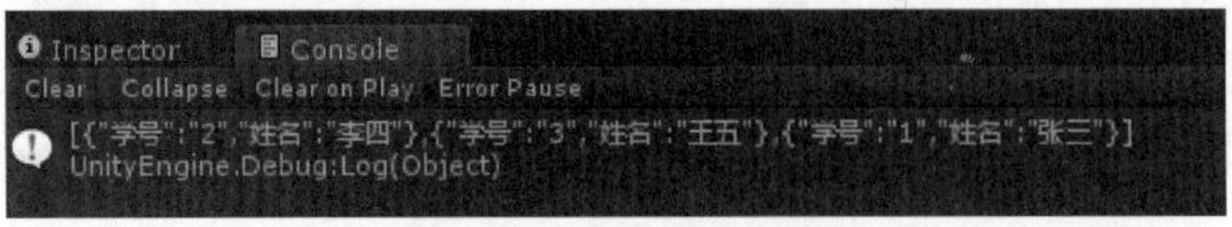

图 5-21 返回的 json 数据

可以看到打印出来是一种带结构的字符串，实际上这是一个 json 结构的键值对结构。这里我需要对 json 结构进行解析。修改 SelectAllInfo()这个函数，代码如下：

```
IEnumerator SelectAllInfo()
{
    yield return StartCoroutine(OperateDatabase.LoadData("select * from 学生信息"));
    infoStr = OperateDatabase.result;
    Debug.Log(infoStr);
    Debug.Log("json 解析：");
    JsonData[] usersInfo = JsonMapper.ToObject<JsonData[]>(infoStr);
    foreach(JsonData jd in usersInfo)
    {
        print(jd["学号"] +" "+jd["姓名"]);
    }

}
```

结果如图 5-22 所示。

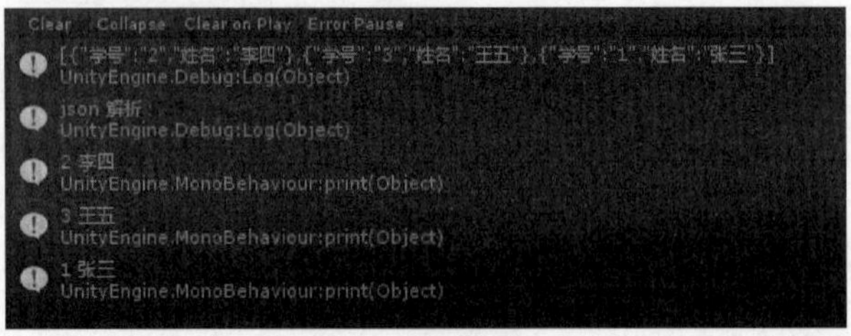

图 5-22 结果显示

在 Unity 中操作 MySQL 数据库时，需要了解以下知识：PHP 连接数据库、json 解析、WWW 和 WWWFrom、SQL 语言等。

OperateDatabase.LoadData(SQL 语句)，只要填入正确的 SQL 语句就可以对数据库进行增删改查。

$db=mysql_connect('localhost','root','qwer1234')，localhost 表示访问本地数据库，可以把它改为其他网址（服务器的 URL）。

json_encode($info); 是对变量进行 json 编码，但是在转换中文的时候会有一些问题，所以会使用下面这段代码让中文正常显示：

$infostr= preg_replace("#\\\u([0-9a-f]{4})#ie", "iconv('UCS-2BE', 'UTF-8', pack('H4', '\\1'))", $infostr);

这里用了 4 个 dll，Unity 调用外包 dll 的时候，要注意把它放到 Plugins 中（没有就新建一个），如图 5-23 所示为该案例 Project 面板截图。

图 5-23 4 个 dll 文件

第 6 章 常用的组件

本章通过介绍自动寻路和地形创建的方法，介绍 Unity 中常用组件的使用技巧。

6.1 导航网格

6.1.1 人物自动寻路到目标点

功能描述：(1) 鼠标单击地面，获取单击的目标点的三维坐标，并且把这个目标点标记出来；(2) 人物自动走到该目标点。

首先创建一个场景，如图 6-1 所示。然后创建一个名为"Ground"的 Layer 层，如图 6-2 所示。

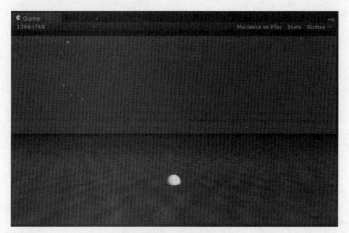

图 6-1 新场景

图 6-2 创建一个 Ground 层

设置场景中的"Plane"，这个物体的 Layer 层为"Ground"，如图 6-3 所示。

图 6-3 设置 Plane 为 Ground 层

创建脚本"Demo1",编辑代码如下:

```csharp
using UnityEngine;
using System.Collections;

public class Demo1 : MonoBehaviour
{

    //将鼠标单击点显示出来,flag 为标记物体
    private GameObject flag;
    void Start ()
    {
        flag = GameObject.Find("Flag");
        //初始时隐藏这个标记
        flag.transform.position = new Vector3(0,-1000,0);
    }


    void Update ()
    {
        //单击鼠标左键发射一条射线与地面相交,设置标记的位置为交点
        if(Input.GetMouseButtonDown(0))
        {
            RayDetectCollision();
        }
    }

//碰撞检测
    void RayDetectCollision()
    {
        Ray ray = Camera.main.ScreenPointToRay(Input.mousePosition);
        RaycastHit hit;
        //将地面的 Layer 层设为第八层对应值为 256
        if(Physics.Raycast(ray,out hit,Mathf.Infinity,256))
        {
            flag.transform.position = hit.point;

        }

    }
}
```

把脚本拖到 Main Camera 这个物体上运行,效果如图 6-4 至图 6-6 所示。

图 6-4　运行效果

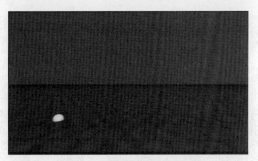
图 6-5　单击地面设置黄球位置

下面来完成第二步：人物能自动走到这个点。首先放 3 个 Cube 作为障碍物，如图 6-7 所示。将整个场景烘培一下，将可以行走的部分与不能行走的部分区别开。

图 6-6　单击地面设置黄球位置

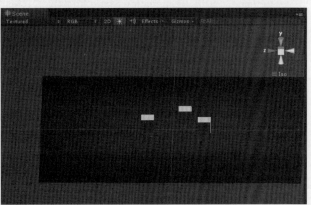
图 6-7　设置静态障碍物

单击"Window→Navigation"，打开导航烘培窗口，如图 6-8、图 6-9 所示。

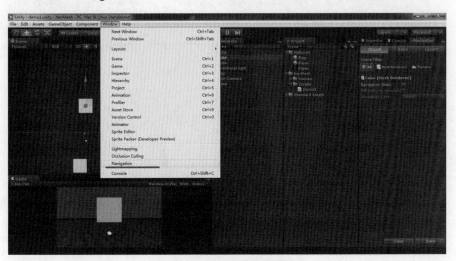
图 6-8　打开网格烘培面板

下面将 3 个 Cube 设置为不能行走，勾选"Navigation-Static"，Navigation-Layer 为"Not Walkable"，设置如图 6-10 所示。

图 6-9　网格烘培面板

图 6-10　对 Cube 进行烘培，设置为不可走

选择 Plane，勾选"Navigation Static"，设置 Navigation Layer 为"Default"，如图 6-11 所示。

图 6-11　设置 Plane 为默认层

单击"Bake"烘培场景，如图 6-12 所示。

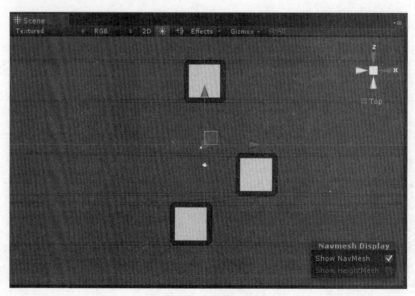

图 6-12　烘培效果

导入带动画的角色，重命名为"Player"，并将人物设为摄影机的父物体，做一个简单的摄影机跟随，如图 6-13 所示。

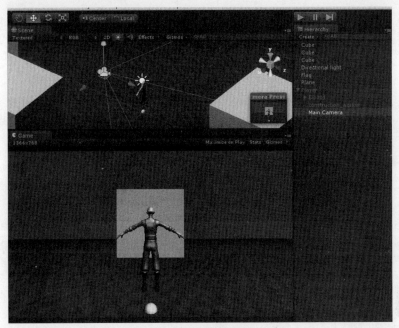

图 6-13 人物导入场景

单击菜单"Component→Navigation→Nav Mesh Agent",给 Player 添加一个名为"Nav Mesh Agent"的组件,如图 6-14 所示。

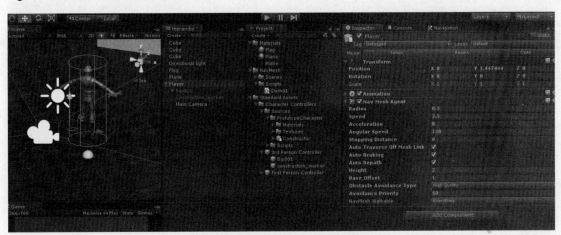

图 6-14 添加导航网格代理

回到 Demo1.cs,编辑脚本如下:

```
using UnityEngine;
using System.Collections;

public class Demo1 : MonoBehaviour
{

//将鼠标单击点显示出来,flag 为标记物体
private GameObject flag;
```

```csharp
//获取导航网格代理组件
private NavMeshAgent agent;

//游戏物体
    private GameObject player;
void Start ()
{
    flag = GameObject.Find("Flag");

    //初始时隐藏这个标记
    flag.transform.position = new Vector3(0,-1000,0);

    agent = GameObject.Find("Player").GetComponent<NavMeshAgent>();

    player = GameObject.Find("Player");
}


void Update ()
{
    //单击鼠标左键发射一条射线与地面相交,设置标记的位置为交点
    if(Input.GetMouseButtonDown(0))
    {
        RayDetectCollision();
    }
    if(flag.transform.position.y > -10)
    {
    int offset_x = (int)Mathf.Abs(player.transform.position.x-flag.transform.position.x);
    int offset_z = (int)Mathf.Abs(player.transform.position.z-flag.transform.position.z);
            //当人物与目标点 x, z 坐标在 0.1 米时, 默认到达目标点
            //如果到达目标点,播放默认动画
            //否则播放行走动画
if(offset_x<0.1f && offset_z<0.1f)
{
    player.animation.CrossFade("idle");
}
else
{
    player.animation.CrossFade("walk");

    }


}


}


void RayDetectCollision()
{
```

```
Ray ray = Camera.main.ScreenPointToRay(Input.mousePosition);
    RaycastHit hit;
    //将地面的 Layer 层设为第八层对应值为 256
    if(Physics.Raycast(ray,out hit,Mathf.Infinity,256))
    {
        flag.transform.position = hit.point;
        //设置人物的目标点
        agent.SetDestination(flag.transform.position);
    }
}
```

这样我们就完成了一个基本的人物自动寻路的功能，人物遇到障碍物会绕过障碍物寻路，效果如图 6-15 所示。

图 6-15　运行效果

6.1.2　导航网格之 Off Mesh Link 使用

Off Mesh Link 分离网格链接分为手动和自动两种。

1. 手动生成分离网格链接

通过一个爬楼梯的案例介绍手动生成分离网格链接的方法。

我们沿用上面一节的脚本（请结合本章对应的工程文件下的 Demo2 场景），搭建如图 6-16 所示的场景。

新建 2 个空物体，在梯子上添加起点和终点，如图 6-17 所示。

给场景中的 Box_Labber 添加 Off Mesh Link 组件，单击 "Component→Navigation→Off Mesh Link"。将起点和终点拖到 Off Mesh Link 中对应的 "Start" 和 "End" 上，如图 6-18 所示。

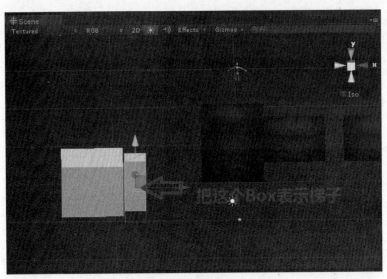

图 6-16 搭建新场景

图 6-17 新场景说明

图 6-18 添加 Off Mesh Link 组件

在 Navigation 面板中设置 Box 和 Box_labber，设置参数如图 6-19 所示。单击"Bake"烘培网格，如图 6-20 所示。

 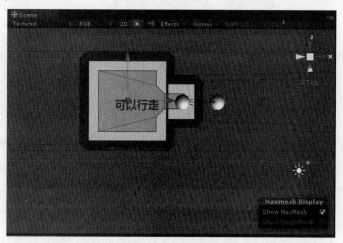

图 6-19　烘培设置　　　　　　　图 6-20　烘培效果

单击 Box 的顶端，就可以看到人物自动上升走到单击的位置。

2. 自动生成分离网格链接

下面通过一个人向下跳和平移的案例介绍自动生成分离网格链接的方法。

搭建一个场景，如图 6-21 所示。

选中这 3 个 Box，在 Navigation 面板设置参数，如图 6-22 所示。

图 6-21　新场景　　　　　　　图 6-22　设置面板

在"Bake"选项卡设置如图 6-23 所示的参数。

- Max Slope：最大倾斜角，超过这个值，人物无法通过。
- Step Height：阻挡的高度，低于这个值，人物能够通过。
- Drop Height：小于该值，人物可以落下。
- Jump Distance：小于该距离，人物可以跳过去。

下面设置 Player 上的 Navmesh Agent 参数，如图 6-24 所示。

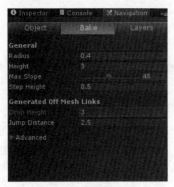

图 6-23 烘培参数

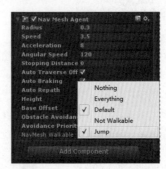

图 6-24 设置可行走层

注意能行走的层要勾选 Jump 层,否则在执行人物落下和平移一段距离时,没有办法得到想要的结果。设置完成以后,重新回到 Navigation 面板,单击"Bake"重新烘培一下场景。烘培完成以后,可以把人物放到一个 Box 上看人物落下,具体执行效果,可以查看本章对应的工程文件 Demo2。

6.1.3 导航网格之动态障碍物 Navmesh Obstacle

新建一个名为 Demo3 的场景,Player 沿用第一个场景代码,场景如图 6-25 所示。

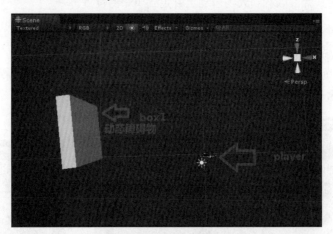

图 6-25 设置动态障碍物

为 Box1 添加组件,单击"Component→Navigation→Nav Mesh Obstacle",Nav Mesh Obstacle 设置如图 6-26 所示。

图 6-26 添加 Nav Mesh Obstacle

为 Box1 添加一段来回移动的代码,具体如下:

```
using UnityEngine;
using System.Collections;

public class Moving : MonoBehaviour
{

    void Start ()
    {

    }


    void Update ()
    {
        //让物体做乒乓来回运动
        transform.position = new Vector3( transform.position.x, transform.position.y, Mathf.PingPong(Time.time, 10));

    }
}
```

这样一个动态的障碍物就添加完成了,完成效果如图 6-27 所示。

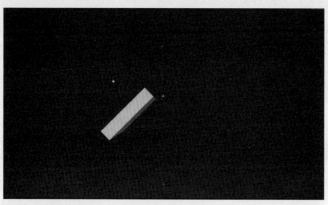

图 6-27 运行效果

6.2 Terrain 地形系统

在 Unity 中包括一个名为 "Terrain" 的地形系统,可以使用它创建一些复杂地形。

单击 "GameObject→Create Other→Terrain" 创建一个地形,如图 6-28 所示。在场景中调整刚创建的 Terrain,如图 6-29 所示。

查看 Inspector 界面,下面是设置地形的工具栏,如图 6-30 所示。

- ![]:表示升高地形,单击它将打开如图 6-31 所示的设置面板。

Unity 游戏开发实用教程

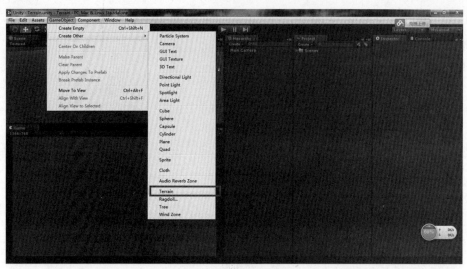

图 6-28 创建地形

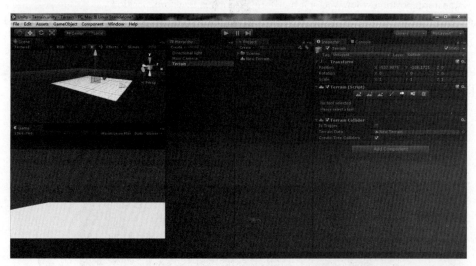

图 6-29 调整地形

图 6-30 地形工具面板

图 6-31 笔刷设置项

单击第一个按钮以后,单击场景内的地形并长按鼠标,可以升高地形;按住 Shift 键再单击鼠标,可以降低地形高度。如图 6-32 所示。

- ：绘制高度工具,可以指定一个高度,然后移动地形上任意部分达到这个高度,当地形达到这个高度以后,它将停止移动并且重置高度,如图 6-33 所示。

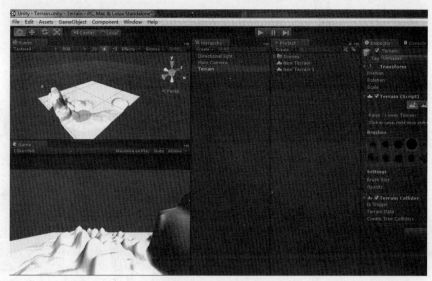

图 6-32 刷出的效果

- ![]：可以使创建的地形高度差变得平滑柔和，如图 6-34、图 6-35 所示为柔化前后对比。

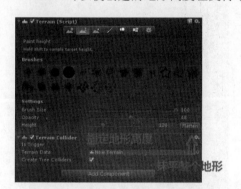

图 6-33 设置项

图 6-34 柔化前效果

- ![]：地形纹理笔刷，可以选择一张贴图纹理，单击笔刷就可以将纹理绘制到地形上。

  单击"Edit Textures→Add Texture"，添加纹理，如图 6-36 所示。

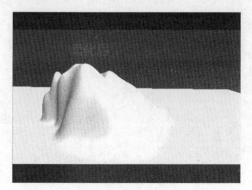

图 6-35 柔化后效果

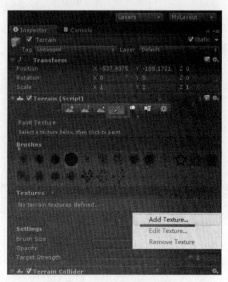

图 6-36 笔刷工具设置面板

单击后会弹出如图 6-37 所示的界面，默认里面是无贴图的。这里导入 Unity 自带的 Terrain 资源包，如图 6-37 所示。

重新打开界面并且设置贴图，如图 6-38、图 6-39 所示。

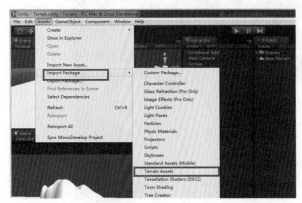

图 6-37　添加 Terrain 资源包

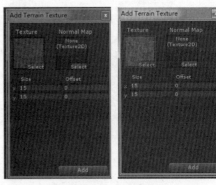

图 6-38　贴图设置

然后回到 Scene，用鼠标单击地形就能绘制纹理了，如图 6-40 所示。

- 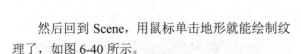：用来绘制树木的笔刷。

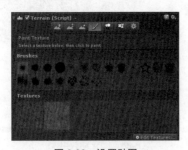

图 6-39　设置贴图

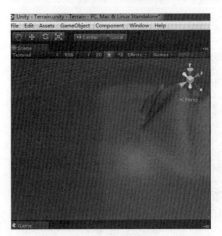

图 6-40　纹理效果

下面在绘制的地形上种一些树木，单击"Edit Trees→Add Tree"，如图 6-41 所示。在弹出的窗口中选择一类树木，如图 6-42 所示。

图 6-41　种树笔刷

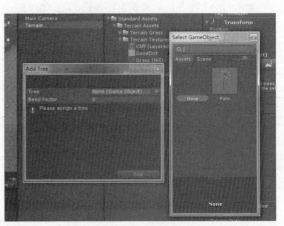

图 6-42　添加树木

双击右边的 Palm 这棵树，单击 Add 将其添加到树木列表中，如图 6-43 所示。

回到 Scene 场景中，使用鼠标左键单击地形就可以添加树林了。如图 6-44 所示为添加树林的地形。

图 6-43　设置

图 6-44　种树效果

- ：可以在地形上绘制一些细节物体，比如草地、岩石等。

下面为绘制的地形添加草丛。单击"Edit-Details→Add Grass Texture"，如图 6-45 所示。弹出如图 6-46 所示的界面，选择一个草地纹理。

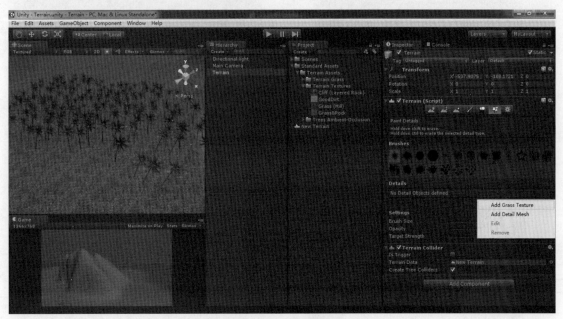

图 6-45　种草笔刷

双击一个草地纹理以后，单击"Add"，如图 6-47 所示。

下面回到 Scene 场景中，使用鼠标左键绘制草地，如图 6-48 所示。

如果想在其他工程文件中使用这个地形，可以把它打包导出，如图 6-49、图 6-50 所示。

图 6-46 添加贴图

图 6-47 添加完成

图 6-48 种草

图 6-49 打包

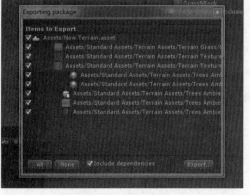

图 6-50 导出面板

第 7 章　多人在线

本章主要介绍多人在线方面的知识，实现多人在线聊天室和多人在线的动画同步。

7.1　开发一个多人聊天室

首先建立一个多人服务器，创建一个名为"Server.cs"的脚本。

```
using UnityEngine;
using System.Collections;

public class Server : MonoBehaviour
{

    //端口号
    private int connectPort = 1990;
    private Vector2 scrollPosition = Vector2.zero;

    private string playerMsgFromDisconnect = "";

    private string msgFromClient = "已经初始化服务器";
    private bool isStartServer = false;
    private bool flag = false;
    void OnGUI ()

    {


        if(isStartServer)
        {

            GUI.Label(new Rect(10,20,400,100),msgFromClient);


            scrollPosition = GUI.BeginScrollView(new Rect(10, 120, 600,480), scrollPosition, new Rect(0, 0, 550, 50*Network.connections.Length));
            for(int i=0;i<Network.connections.Length;i++)
            {
                GUI.Label(new Rect(10,45*i+10,500,40),Network.connections[i].ipAddress+"已连接！");
            }
            GUI.EndScrollView();


        }
        else
        {
```

```
                    //初始化多人在线服务器
                    if(GUI.Button(new Rect(10,10,100,60),"创建服务器"))
                    {
                            Network.InitializeServer(32, connectPort, false);
                            isStartServer = true;
                    }
            }
}

//当用户从服务器断开，在服务器调用这个函数
void OnPlayerDisconnected(NetworkPlayer player)
{
    playerMsgFromDisconnect = player.ipAddress+" 断开连接！";
    msgFromClient = playerMsgFromDisconnect;
    Network.RemoveRPCs(player);
    Network.DestroyPlayerObjects(player);
}

//接收请求的方法，注意要在上面添加[RPC]
[RPC]
void ReciveMessage(string msg, NetworkMessageInfo info)
{
//刚从网络接收的数据的相关信息，会被保存到 NetworkMessageInfo 这个结构中
    msgFromClient = info.sender.ipAddress +"用户："+msg;
}
}
```

下面介绍其中的一些重要函数与类：

（1）Static Function InitializeServer (connections : int, listenPort : int, useNat : bool) : NetworkConnection-Error

connections 是允许的入站连接或玩家的数量，listenPort 是要监听的端口，useNat 设置 Nat 穿透功能。

（2）Network.InitializeServer(32, connectPort, false)

这句代码是用来创建服务器，32 表示可以连接 32 个用户，connectPort 表示端口号，false 表示不穿透外网，IP 地址是内网地址。

（3）Network.connections：这个数组里有所有连接的玩家。

（4）Function OnPlayerDisconnected (player : NetworkPlayer) : void

每当一个玩家从服务器断开时,在服务器调用这个函数。一般要在这个函数里面清理断开玩家遗留的痕迹。

(5)[RPC]

Void ReciveMessage(string msg, NetworkMessageInfo info)是一个 RPC 调用函数,这个函数在所有连接端调用;要使用 RPC 调用函数,必须要给游戏物体添加 NetworkView 组件。

在场景中创建一个空物体,重命名为"Server",将 Server.cs 添加到 Server 物体上,然后添加单击"Component→Miscellaneous→Network View"组件,参数默认。如图 7-1 所示。

图 7-1 设置面板

下面建立一个客户端,代码如下:

```
using UnityEngine;
using System.Collections;

public class Client : MonoBehaviour {

    private string ip = "127.0.0.1";
    private int port = 1990;
    private string msg = "";
    private string otherMsg = "";
    void Start ()
    {

    }


    void OnGUI ()
    {

        switch(Network.peerType)
        {
            //禁止客户端连接运行,服务器未初始化
            case NetworkPeerType.Disconnected:
                if (GUI.Button(new Rect(10,10,100,40),"连接服务器"))
                {
                    NetworkConnectionError error = Network.Connect(ip,port);
```

```
            }
            break;
            //运行于服务器端
        case NetworkPeerType.Server:
            break;
            //运行于客户端
        case NetworkPeerType.Client:
            msg = GUI.TextArea(new Rect(10, 10, 400, 300), msg, 300);
            if (GUI.Button(new Rect(10,320,100,40),"发送信息"))
            {
                //发送给接收的函数，模式为全部，参数为信息
                networkView.RPC("ReciveMessage", RPCMode.Others, msg);
                msg = "";
            }
otherMsg = GUI.TextArea(new Rect(430, 10, 400, 300), otherMsg, 300);
            break;
            //正在尝试连接到服务器
        case NetworkPeerType.Connecting:
            break;
        }
    }
    //接收请求的方法，注意要在上面添加[RPC]
    [RPC]
    void ReciveMessage(string msg, NetworkMessageInfo info)
    {
    //刚从网络接收的数据的相关信息，会被保存到NetworkMessageInfo这个结构中
        otherMsg = info.sender.ipAddress +"用户："+msg;;
    }
}
```

下面介绍客户端中的重要函数：

（1）static function Connect (IP : string, remotePort : int, password : string = "") : Network ConnectionError

连接到服务器，第一个参数是 IP 地址，第二个是端口号，第三个是密码。这里只设置前 2 个参数。IP 地址相当于家的地址，端口号相当于家的大门钥匙，大多数 TCP/IP 可以给临时端口分配 1024~5000 之间的端口号。注意 IP 地址为服务器地址，端口号要和服务器一样。

（2）networkView.RPC("ReciveMessage", RPCMode.Others, msg)

通过 networkView 这个组件调用 RPC 函数。发送消息给其他客户端和服务器，消息内容为 msg。RPCMode 名为远程过程调用模式，有 5 个值：Server 表明消息仅发送给服务器；Others 表明消息发送给其他人；OthersBuffered 表明消息除了自己之外，发送到每个人，并添加到缓冲区；All 表明发送给所有人；AllBuffered 表明将消息发送给每个人，并且消息放到缓冲区。

下面我们新建一个场景，在场景中新建一个空物体名为"Client"，将脚本添加到 Client，并且添加 Network View 组件，如图 7-2 所示。保存场景。然后将建立的 Server 和 Client 2 个场景分别发布，注意不要一起发布。

先发布 Server 场景，如图 7-3 所示。

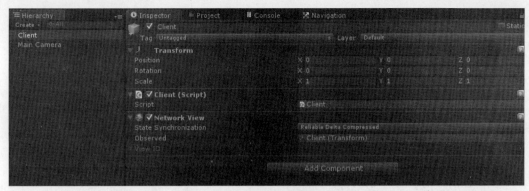

图 7-2 客户端设置面板

设置 Player，如图 7-4 所示（注意一定要勾选 Run In Background，表示程序可以在后台运行）。

图 7-3 发布 Server

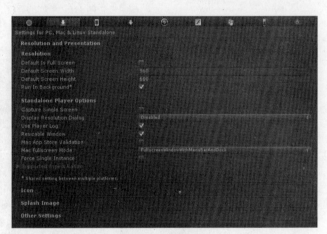

图 7-4 发布设置

同样发布 Client，如图 7-5 所示。单击打开服务器，如图 7-6 所示。

图 7-5 发布客户端

图 7-6 服务器打开效果

打开客户端，并且发送一段消息，如图 7-7 所示。

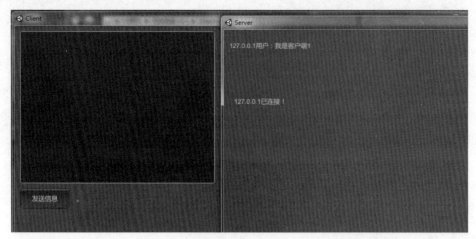

图 7-7　客户端运行

我们再打开一个客户端，如图 7-8 所示。

图 7-8　完成效果

7.2　动画同步与位置同步

下面通过介绍多人在线过程中动画同步来进一步学习 RPC 调用。整个工作流程为：首先创建服务器，服务器初始化成功后，等待客户端连入；当一个客户端打开以后，通过 IP 和端口号与服务器连接，创建一个 Player（人物角色）并且要把其他客户端的人物角色复制出来，自己控制的人物角色的位置和动画能够在其他客户端同步。

（1）首先创建一个服务器

创建一个空物体，名为"NetworkProcess"，给这个空物体添加组件"Component→Miscellaneous→Network View"，并且将创建好的 Server.cs 脚本拖到这个物体上，如图 7-9 所示。

第 7 章 多人在线

图 7-9 设置面板，添加 Network View 组件

Server.cs 脚本如下：

```
using UnityEngine;
using System.Collections;

public class SeverCSharp : MonoBehaviour
{
    private int connectPort=25001;

    void Start()
    {

    }
    void OnGUI ()
    {
        if (Network.peerType == NetworkPeerType.Disconnected)
        {
            Network.InitializeServer(32, connectPort, false);
        }
    }
    void OnPlayerDisconnected(NetworkPlayer player)
    {
        Debug.Log(player.ipAddress+"断开");
        Network.RemoveRPCs(player);
        Network.DestroyPlayerObjects(player);
    }


    //同步调用默认动画
    [RPC]
    public void SynIdle ()
    {

    }
```

```
//同步调用行走动画
[RPC]
void    SynWalk ()
{

}

}
```

脚本中的大部分内容与上一个聊天室的脚本类似，这里多了 2 个 RPC 函数，函数内容为空，这是因为 RPC 调用的时候选择所有人都执行，如果没有这 2 个函数会报错。最后，可以把场景内其他游戏物体去掉，只保留 "NetworkProcess"，将服务器端发布。发布设置如图 7-10 所示，注意要保证服务器后台也能运行。

单击运行 Server.exe，如图 7-11 所示。

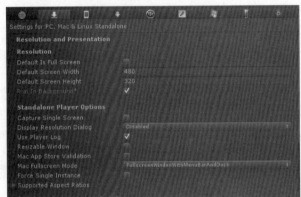

图 7-10 保证服务器在后台也能运行

图 7-11 运行效果

（2）制作客户端

创建一个名为 "Client.cs" 的脚本，把这个脚本拖到 NetworkProcess 物体上（去掉上一步 Server 脚本），如图 7-12 所示。

图 7-12 添加客户端脚本

打开"Client.cs",编辑脚本如下:

```csharp
using UnityEngine;
using System.Collections;

public class Clicent : MonoBehaviour
{

    private string severIP;//服务器 IP 地址
    private int severPort;//服务器端口号
    public GameObject playerPrf;//人物模型
    void Start ()
    {
        severIP="127.0.0.1";
        severPort = 25001;
        if(Network.peerType == NetworkPeerType.Disconnected)
        {
            Network.Connect(severIP,severPort);
        }
    }

    void OnConnectedToServer()
    {
        Debug.Log("客户端成功连接");
        Network.Instantiate(playerPrf,new Vector3(0,10,0),Quaternion.identity,0);
    }

    void OnDisconnectedFromServer(NetworkDisconnection info)
    {
        Debug.Log("客户端断开连接");

    }

}
```

其中,Network.Connect(severIP,severPort)表示通过 IP 地址和端口号连接服务器;void OnConnectedToServer()这个函数会在客户端连接到服务器后自动调用;Network.Instantiate(playerPrf,new Vector3(0,10,0),Quaternion.identity,0)表示当客户端连接到服务器以后实例化一个 PlayerPrf 游戏物体,三维坐标为(0,10,0),无旋转,网络组为 0。

接下来单击导入"Assets→Import Package→Character Controller"这个资源包,在这个场景中创建一个地形,添加草地和树木,将 Character Controller 中的 Constructor 文件拖到场景中,重命名为 Player,并且添加"Component→Physics→Character Controller"到 Player 上,如图 7-13 所示。

图 7-13 Player 设置面板

（3）下面建立一个人物位移控制脚本 CharacterMovement.cs，并且将脚本拖到 Player 上，脚本如下：

```
using UnityEngine;
using System.Collections;
public class CharacterMovement : MonoBehaviour
{
    //重力影响
    private float gravity = 10;
    //是否与地面碰撞
    private bool   grounded = false;
    //移动的向量
    private Vector3 moveDirection = Vector3.zero;
    //移动速度
    private   float moveSpeed = 5f;
    //角色控制器
    private    CharacterController controller;


    void   Start ()
    {
        controller = GetComponent<CharacterController>();
    }

    void   Update ()
    {
        if(grounded)
        {
            moveDirection = new Vector3(Input.GetAxis("Horizontal"),0,Input.GetAxis("Vertical"));
            moveDirection = transform.TransformDirection(moveDirection);
```

```
            moveDirection *= moveSpeed;

    }


        transform.rotation = Quaternion.Euler(0,Camera.main.transform.eulerAngles.y,0);
            //重力影响，人物下降
      moveDirection.y -= gravity * Time.deltaTime;


        if(controller.active)
        {
            //人物移动
            CollisionFlags flags= controller.Move(moveDirection * Time.deltaTime);
                //判断是否与地面发生碰撞
            grounded = (flags & CollisionFlags.Below) != 0;
        }



  }
}
```

这个脚本主要控制人物前后左右移动，可以按 W、A、S、D 和箭头键（方向键）来实现，但是可以发现移动的时候人物没有运动动画，显得很不自然，所以加入下面这段代码。

```
if(Input.GetAxis("Horizontal")!=0 || Input.GetAxis("Vertical")!=0)
{

//按住 WASD 调用前后左右移动动画

    if(Input.GetAxis("Vertical") > 0)
    {
        networkView.RPC("SynWalk",RPCMode.All);
    }
    if(Input.GetAxis("Vertical") < 0)
    {
        networkView.RPC("SynWalk",RPCMode.All);
    }

    if(Input.GetAxis("Horizontal") > 0)
    {
        networkView.RPC("SynWalk",RPCMode.All);
    }
if(Input.GetAxis("Horizontal") < 0)
    {
        networkView.RPC("SynWalk",RPCMode.All);
    }
}
//没有按前后左右控制按钮时，人物调用默认动画
else if(Input.GetAxis("Horizontal")==0 && Input.GetAxis("Vertical")==0)
```

```
{
    networkView.RPC("SynIdle",RPCMode.All);
}
```
这 2 个 RPC 函数放在 AnimationController.cs 脚本里，脚本如下：
```
using UnityEngine;
using System.Collections;

public class AnimationController : MonoBehaviour
{

    //同步调用默认动画
    [RPC]
    public void   SynIdle ()
    {
        animation.CrossFade("idle");
    }

    //同步调用行走动画
    [RPC]
    void   SynWalk ()
    {
        animation.CrossFade("walk");
    }

}
```

现在还有一个问题，当实例化几个客户端的 Player 后，在前进后退的时候其他客户端的 Player 也会同样受到这个脚本的影响，所以在实例化 Player 的时候要把 CharacterMovement 和 Character Controller 这 2 个组件删除，代码如下：

```
//当复制游戏物体的时候
void OnNetworkInstantiate(NetworkMessageInfo msg)
{
//复制的是自己
if(networkView.isMine)
    {
        gameObject.name = "Player";
        Camera.main.GetComponent<MainCameraController>().target = gameObject.transform;
    }
    //复制其他客户端的游戏物体
    else
    {
        if(gameObject.GetComponent<CharacterMovement>() != null)
        {
            Destroy(gameObject.GetComponent<CharacterMovement>());
        }

        if(gameObject.GetComponent<CharacterController>() != null)
        {
            Destroy( gameObject.GetComponent<CharacterController>());
        }
```

}

}

其中，void OnNetworkInstantiate(NetworkMessageInfo-msg)表示当使用 Network.Instantiate 实例化 Player 后，Player 上会自动调用这个函数。NetworkView.isMine 表示这个物体是你自己控制的而不是其他客户端控制的。给 Player 也添加上 "Component→Miscellaneous→Network View" 组件，参数默认。Player 上面的组件如图 7-14 所示。

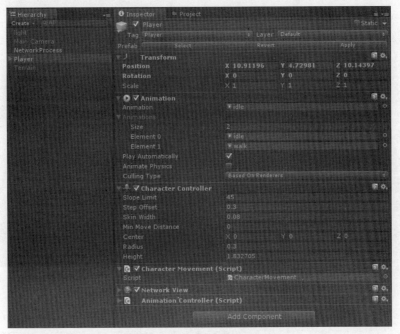

图 7-14　添加 Network View 组件

检查 Player 是否都添加了这些组件或者脚本（其实脚本也是组件之一），将 Player 拖到 Project→Models 中变成 Prefab，将这个 Prefab 拖到 NetworkProcess→Client→Player Prf 上，最后删除 Hierarchy 中的 Player。如图 7-15 所示。

图 7-15　Client 脚本设置

将客户端发布出来，发布的 EXE 如图 7-16 所示。

图 7-16　发布客户端

单击打开 server.exe，然后打开 client.exe，如图 7-17 所示为客户端的效果图。
如果出现这种情况，那么需要重新发布一下服务器端设置，如图 7-18 所示。

图 7-17　运行效果

图 7-18　重新发布选项

将 Client.cs 也放到服务器上，但是不激活它，重新发布，效果如图 7-19 所示。

图 7-19　完成图

第 8 章　基于 Unity 的安卓开发

本章主要介绍基于 Unity 的安卓开发，主要包括关于 xml 的读写、视频的播放以及其他在安卓平台上需要注意的问题。

8.1　安卓开发环境配置

8.1.1　安装 jre

jre(java runtime environment)，就是 java 程序的运行环境。首先到网上下载一个 jre 安装，下载地址为 http://www.java.com/en/download/manual.jsp#win，如图 8-1 所示，可以根据计算机的位数选择版本。

下载完成以后双击 jre-7u51-windows-i586.exe 安装，如图 8-2、图 8-3 所示。

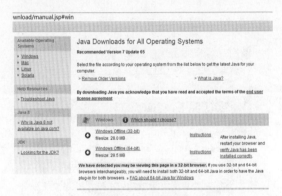

图 8-1　jre 下载页面

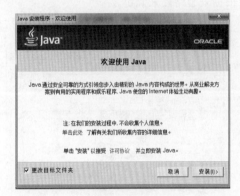

图 8-2　安装

可以默认安装。安装完成以后，需要配置环境变量，单击"计算机->属性"，找到"高级系统设置"，如图 8-4 所示。

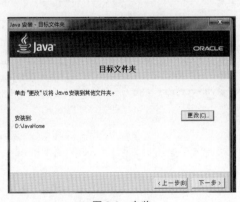

图 8-3　安装

图 8-4　设置

然后单击"环境变量",如图 8-5 所示。在"系统变量"里面新建一个名为"java_home"的文件,如图 8-6 所示。

图 8-5 设置

图 8-6 设置系统变量

单击"确定"按钮,再新建一个名为"classpath"的系统变量,如图 8-7 所示。classpath 的值要根据安装路径来确定。

下面在环境变量里面找到"Path",对"Path"进行配置,如图 8-8 所示。

图 8-7 设置 Classpath 的值

图 8-8 设置 Path

一般情况下"Path"这个变量已经存在,只要把";D:\JavaHome\bin"(安装目录)填进去就可以了,注意要用分号";"与前后分隔开。下面可以测试是否安装成功。

在"运行"中输入"cmd"打开控制台,在控制台中输入"java -version",特别注意 java 和-version 之间有一个空格,按"Enter"键。如图 8-9 所示。

第 8 章 基于 unity 的安卓开发

图 8-9 验证是否安装成功

当控制台出现类似信息时，表明 jre 已经安装成功了。

8.1.2 下载更新 android SDK

可以在官网或者其他提供 SDK 包下载的网上下载 android SDK，下载完以后如图 8-10 所示。

对系统变量 Path 进行配置，将安卓 SDK 路径添加进去，如图 8-11 所示。

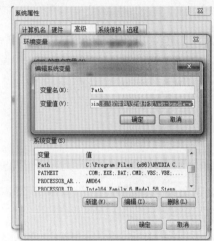

图 8-10 下载的 Sdk 包　　　　　　　　图 8-11 设置

打开 eclipse.exe，单击"Andriod SDK Manager"，如图 8-12 所示。

图 8-12 SDK 管理面板

单击"Android SDK Manager->Tools->Options",参数设置如图 8-13 所示。

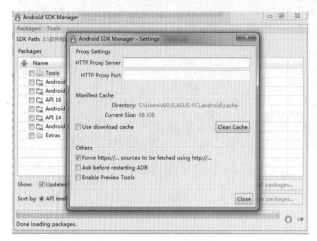

图 8-13 设置

关闭以后,勾选需要更新的版本,如图 8-14 所示。

注意尽量将各个 SDK 都更新完。更新完毕后,打开 Unity 新建一个名为"Hello World"场景,然后切换平台到安卓,如图 8-15 所示。

图 8-14 更新　　　　　　　　　图 8-15 切换到安卓发布平台

创建一个名为"HelloWorld.cs"的脚本:

```
using UnityEngine;
using System.Collections;

public class HelloWorld : MonoBehaviour
{
    bool isShow = false;
```

```
void Start ()
{

}


void OnGUI ()
{
    if(isShow)
    {
        GUI.Label(new Rect(10,30,200,60),"Hello World!");
    }

    if(GUI.Button(new Rect(30,100,100,40),"click"))
    {
        isShow = !isShow;
    }
}
}
```

设置屏幕尺寸为 320×480，如图 8-16 所示。

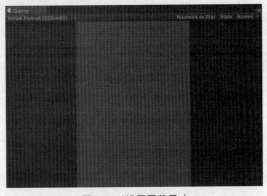

图 8-16　设置屏幕尺寸

单击"Unity Edit -> Preferences."，设置如图 8-17 所示。

图 8-17　设置 SDK 路径

现在发布产品，发布设置如图 8-18 所示。

图 8-18　发布设置

产品图标可以自定义，默认是 Unity 的标志。下面设置的是屏幕可以自动选择，如图 8-19 所示。

图 8-19　发布设置

本例选择发布的安卓版本为 4.0，同时要注意修改 Bundle Identifier* 的值，原来"com.Company.ProductName"中的"Company"要改掉。如图 8-20 所示。

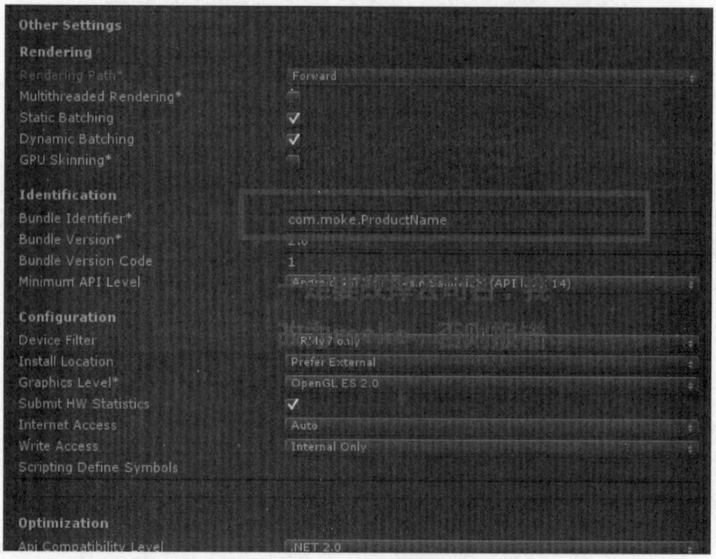

图 8-20　发布设置

接下来设置过渡画面，如图 8-21 所示。最后点击发布，如图 8-22 所示。

图 8-21　设置过渡画面

图 8-22　发布

如图 8-23 所示是发布的 apk，拷贝到安卓手机里安装就可以了。

图 8-23　发布的 apk

8.2　简单的触屏操作示例

8.2.1　单指旋转物体

在场景中创建一个 box，设置屏幕尺寸为 320×480，如图 8-24 所示。

创建脚本"TouchDemo.cs"，编辑脚本如下：

图 8-24　Cube

```
using UnityEngine;
using System.Collections;

public class TouchDemo : MonoBehaviour
{

    public float speed = 0.01F;

    void Update ()
    {
```

```csharp
            if (Input.touchCount > 0 && Input.GetTouch(0).phase == TouchPhase.Moved)
            {
                    Vector2 touchDeltaPosition = Input.GetTouch(0).deltaPosition;
                    //垂直方向的位移大于水平方向位移，则上下旋转
                    if(Mathf.Abs(touchDeltaPosition.y)>Mathf.Abs(touchDeltaPosition.x))
                    {
                            transform.Rotate(new Vector3(-touchDeltaPosition.y,0,0),Space.World);
                    }
                    else
                    {
                            //垂直方向的位移小于水平方向位移，则左右旋转
                            transform.Rotate(new Vector3(0,touchDeltaPosition.x,0),Space.World);
                    }



            }


    }

}
```

当用手指在屏幕上滑动的时候，会使物体进行水平或者垂直旋转。

8.2.2 多点缩放物体

本例主要通过两个手指触摸屏幕来进行缩放物体。在场景中创建一个 box，创建一个名为"MultiPointTouchDemo.cs"的脚本，编辑脚本如下：

```csharp
using UnityEngine;
using System.Collections;

public class MultiPointTouchDemo : MonoBehaviour {

    private float lastDist = 0;

    private float curDist = 0;
    private float distance = 0;
    private GameObject mainCamera;
    void Start ()
    {
        mainCamera = GameObject.Find("Main Camera");
    }


    void Update ()
    {
        if (Input.touchCount > 1 && (Input.GetTouch(0).phase == TouchPhase.Moved || Input.GetTouch(1).phase == TouchPhase.Moved))
        {
            Touch touch1 = Input.GetTouch(0);
```

```
                    Touch touch2 = Input.GetTouch(1);
                    float curDist = Vector2.Distance(touch1.position, touch2.position);

                    if(curDist > lastDist)
                    {
                            mainCamera.transform.position += new Vector3(0,0,0.1f);
                    }
                    else
                    {
                            mainCamera.transform.position += new Vector3(0,0,-0.1f);
                    }
                    lastDist = curDist;
            }
    }
    void OnGUI()
    {
            if(GUI.Button(new Rect(10,400,300,60),"Quit"))
            {
                    Application.Quit();
            }
    }
}
```

脚本通过判断前后 2 帧的两点间的距离来控制物体离摄影机的远近，从而实现物体缩放（实际上物体没有缩放）。两点间如果后一帧的距离比前一帧的距离大就拉近，否则拉远。

8.3 在安卓上操作 Xml

8.3.1 安卓上如何读取 Xml

前面章节中介绍的读取 Xml 的方式在安卓上有些不同。在安卓上由于路径问题可以不能使用 Application.dataPath。作为代替，可以使用 Application.persistentDataPath。实际上除了这点不同以外，一般读取 Xml 的方式在安卓上也是可行的。

要读取 Xml，首先建立一个 test.txt，把它放到 Resources 文件夹下，如图 8-25 所示。

将想读取的 Xml 内容放到 test.txt 文件中，下面是编辑的 Xml：

```
<?xml version="1.0" encoding="UTF-8" standalone="no"?>
<root>
<child>安卓简单读取 xml</child>
</root>
```

图 8-25　Resources 下的 test.xml

方法很简单，先把 txt 文件读取到内存里，然后通过 LoadXml 这种方式加载。创建一个脚本名为 "ReadXmlDemo.cs"，编辑脚本如下：

```
using UnityEngine;
using System.Collections;
using System.Xml;
public class ReadXmlDemo : MonoBehaviour
{

    private TextAsset    textAsset;
    private XmlNode root;
    void Start ()
    {
        //导入文本
        textAsset = Resources.Load("test") as TextAsset;
        XmlDocument xmlDoc = new XmlDocument();
        //加载 Xml
        xmlDoc.LoadXml(textAsset.text);
        root = xmlDoc.DocumentElement;

    }


    void OnGUI ()
    {
        GUI.Label(new Rect(10,100,200,50),root.SelectSingleNode("child").InnerText);

        if(GUI.Button(new Rect(10,200,100,60),"退出"))
        {
            Application.Quit();
        }
    }
}
```

这样一个简单的基于安卓上读取 Xml 的方式就完成了，关键是 LoadXml()这个函数参数为文本，所以只要能将 Xml 文本加载到内存再使用这个函数，那么读取 Xml 就不成问题了。

最后 Unity 中的效果如图 8-26 所示。

8.3.2　安卓上如何写入 Xml

先保存一个 Xml 的框架到安卓下的某个路径，然后获取该路径下的 Xml，进行读写。

同样创建一个名为"WriteXmlDemo.cs"的脚本，编辑脚本如下：

图 8-26　完成效果

```
using UnityEngine;
using System.Collections;
using System.IO;
using System.Xml;
public class WriteXmlDemo : MonoBehaviour
{
```

```csharp
    private string xml;
    private string stringToEdit = "";
    private XmlDocument xmlDoc;
    void Start ()
    {
//简单的 Xml 模板
xml = "<?xml version=\"1.0\" encoding=\"UTF-8\" standalone=\"no\"?><root>熊出没</root>";
        if(!File.Exists(Application.persistentDataPath+"/myXml.xml"))
        {
             xmlDoc = new XmlDocument();
            xmlDoc.LoadXml(xml);
            xmlDoc.Save(Application.persistentDataPath+"/myXml.xml");
        }
        else
        {
            xmlDoc = new XmlDocument();
            xmlDoc.Load(Application.persistentDataPath+"/myXml.xml");
            stringToEdit = xmlDoc.DocumentElement.InnerText;
        }

    }


    void OnGUI ()
    {
        //修改 Xml 内容
        stringToEdit = GUI.TextField (new Rect(10, 10, 200, 60), stringToEdit, 25);
        if(GUI.Button(new Rect(50,100,100,60),"保存"))
        {
            xmlDoc.DocumentElement.InnerText = stringToEdit;
            xmlDoc.Save(Application.persistentDataPath+"/myXml.xml");
            stringToEdit = "";
        }

        if(GUI.Button(new Rect(50,200,100,60),"退出"))
        {
            Application.Quit();
        }

    }
}
```

　　上面这段代码的意思是打开软件时，首先判断在 Application.persistentDataPath 下的"myXml.xml"这个文件是否存在，如果不存在就导入 Xml 模板在该路径下创建一个；如果存在就读取根节点的内容显示出来。同时可以在文本输入框里输入文本来修改根节点的内容并保存。下次打开的时候就可以看到上一次编辑的内容了。

8.4　安卓上播放视频

　　前面的章节中介绍了使用 MovieTexture 播放控制视频的方法，但是移动端是不支持

MovieTexture 的，Unity 提供了一个新的类 iPhoneUtils 来控制在移动平台的视频播放。可以使用 iPhoneUtils.PlayMovie()这个函数播放本地视频文件，在 Resources 文件夹下创建一个名为 Streaming Assets 的文件夹，在里面存放想要播放的视频，如图 8-27 所示，在 StreamingAssets 下有一个名为 XiaomiPhone.mp4 的视频文件。

图 8-27　StreamingAssets 下的视频文件

创建一个名为"PlayMovieDemo.cs"的脚本，脚本如下：

```
using UnityEngine;
using System.Collections;

public class PlayMovieDemo : MonoBehaviour
{

    void OnGUI ()
    {
        if(GUI.Button(new Rect(5,100,310,60),"Play Movie"))
        {
            //播放 StreamingAssets 文件夹下 XiaomiPhone.mp4 这个文件;
            //背景是白色
            //点击一下视频结束播放
            iPhoneUtils.PlayMovie("XiaomiPhone.mp4",Color.white,iPhoneMovieControlMode.CancelOnTouch);

        }

        //退出
        if(GUI.Button(new Rect(5,220,310,60),"Quit"))
        {
            Application.Quit();

        }
    }
}
```

脚本很简单，主要就是一个函数 iPhoneUtils.PlayMovie，调用这个函数时在播放影片期间会暂停 Unity 应用，播放完成后会恢复 Unity 应用。其主要支持的常用视频格式有.mp4、.mov、.mpv、.3gp 等。下面是 PlayMovie 完整的参数列表：

PlayMovie (path : string, bgColor : Color, controlMode : iPhoneMovieControlMode = iPhoneMovieControlMode.Full, scalingMode : iPhoneMovieScalingMode = iPhoneMovieScalingMode.AspectFit) : void

第一个参数为视频路径；第二个参数为播放器指定背景色；第三个参数是影片控制模式；第四个参数为影片的缩放模式。

影片的控制模式有 4 种，说明如下：

iPhoneMovieControlMode.Full：显示控制影片播放的完整控制功能。包含播放/暂停，音量和时间线控件。
iPhoneMovieControlMode.Minimal ：控制功能最少。
iPhoneMovieControlMode.CancelOnTouch：用户单击正在播放的视频结束视频播放。
iPhoneMovieControlMode.Hidden：不显示任何控件。

影片缩放模式也有 4 种，说明如下：

iPhoneMovieScalingMode .None：无缩放。
iPhoneMovieScalingMode.AspectFit：同比例缩放。
iPhoneMovieScalingMode.AspectFill：同比例缩放视频直到填满整个屏幕。
iPhoneMovieScalingMode.Fill：不保证高宽比的缩放填充整个屏幕。

下面开始发布，发布设置如图 8-28、图 8-29 所示。

图 8-28 发布设置

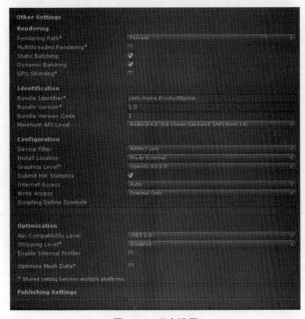

图 8-29 发布设置

发布完成后得到一个 apk 文件，如图 8-30 所示。

图 8-30 发布得到的 apk 文件

除了上面这种方法外，也可以用 Handheld.PlayFullScreenMovie 这个函数来播放视频，代码如下：

Handheld.PlayFullScreenMovie("XiaomiPhone.mp4",Color.black,FullScreenMovieControlMode.CancelOnInput);

视频需要放到 StreamingAssets 下，发布设置不变，发布即可。

第 9 章 常见问题、错误及插件介绍

9.1 常见问题

9.1.1 js 脚本如何与 C#互相调用

Unity 可以创建多种语言类型的脚本文件，常用的就是 js 和 C#。在实际工作中，并不赞同使用多种语言混合编码，个人比较支持统一使用 C#。然而有的时候你可能导入了一些插件，在这些插件只有 js 的时候，就需要了解 js 脚本和 C#互相调用的方法。

先介绍 js 如何调用 C#，js 的脚本如下：

```
#pragma strict
var script : CSharpTest;
function Start ()
{
    script = gameObject.GetComponent("CSharpTest");
    script.Say();

}

function Say ()
{
    print("CS Call JS");
}
```

被调用的 C#如下：

```
using UnityEngine;
using System.Collections;

public class CSharpTest : MonoBehaviour
{

    public     void Say()
    {
        Debug.Log("js Call C#");
    }

}
```

这两个脚本建立以后，将 C#脚本放到 Standard Assets 文件夹下，js 脚本放到 Scripts 下，如图 9-1 所示。

单击运行，结果如图 9-2 所示。

图 9-1 将 2 个脚本放在不同的文件目录下

图 9-2 完成效果

接下来介绍 C#调用 js 脚本的方法。

重新编辑 C#脚本，代码如下：

```csharp
using UnityEngine;
using System.Collections;
public class CSharpTest : MonoBehaviour
{
    void Start()
    {
        gameObject.GetComponent<JsTest>().Say();
    }

    public       void Say()
    {
        Debug.Log("js Call C#");
    }
}
```

编辑过的 js 脚本如下：

```
#pragma strict
//var script : CSharpTest;
function Start ()
{
//    script = gameObject.GetComponent("CSharpTest");
//    script.Say();

}

function Say ()
{
    print("CS Call JS");
}
```

然后更改两个脚本所在文件，将 CSharpTest.cs 放到 Scripts 下，将 jsTest 放到 Standard Assets 目录，如图 9-3 所示。运行结果如图 9-4 所示。

调用哪个脚本，哪个脚本文件就要放在"Standard Assets"、"Pro Standard Assets"和"Plugins"这三个目录中的任何一个中，因为这三个文件最先被编译，后编译的类才能找到先编译的类，所以 C#与 js 文件不要放在一个目录中。例如，使用 js 调用 C#，如果放在同一个目录中就会得到下面的错误，如图 9-5 所示。

图 9-3 文件目录

图 9-4 完成效果

图 9-5 错误提示

9.1.2 Unity 脚本如何与网页脚本互相调用

本例是让网页发送信息给 Unity，显示一条信息在 Unity Player 上，然后 Unity 里面发送一条消息到网页。

先创建一个名为 "UnitySendMsgToHtml.cs" 的脚本，编辑脚本如下：

```
using UnityEngine;
using System.Collections;

public class UnitySendMsgToHtml : MonoBehaviour
{
    private string    msg = "";

    void OnGUI ()
    {
        if(GUI.Button(new Rect(0,0,200,70),"发送消息给网页"))
        {
            SendMsgToHtml();
        }

        if(msg.Length>0)
            GUI.Label(new Rect(0,100,200,60),msg);
    }

     void SendMsgToHtml()
    {
        Application.ExternalCall("MyFunction", "msg come from Unity!");
    }

    public void ReceiveMsgFromHtml(string msg)
    {
        this.msg = msg;
    }
}
```

这个脚本中有两个关键函数：SendMsgToHtml 向网页发送消息；ReceiveMsgFromHtml 接受来自网页的消息。

Application.ExternalCall()可以用来调用网页上的函数，其第一个参数是函数名，后面有多个函数的参数列表，这个函数大家可以参考 Unity 帮助文件学习。

下面将脚本拖到 Main Camera 这个游戏物体上，如图 9-6 所示。然后设置发布参数，如图 9-7 所示。

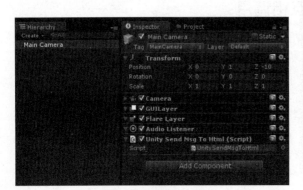

图 9-6　发送消息到网页上

图 9-7　发布为 Web Player

发布完成以后，找到发布的网页，如图 9-8 所示。

图 9-8　发布的网页

打开网页，添加如下代码：

```
Function MyFunction（arg）
{
    alert（arg）;
}

var u = new UnityObject2（）;
u.initPlugin（jQuery（"#UnityPlayer"）[0], "Example.Unity3d"）;
function SendMsgTonUnity（）
{
    u.getUnity（）.SendMessage（"Main Camera", "ReceiveMsgFromHtml", "msg come from html"）;
}
```

最后看一下运行效果，如图 9-9 所示。

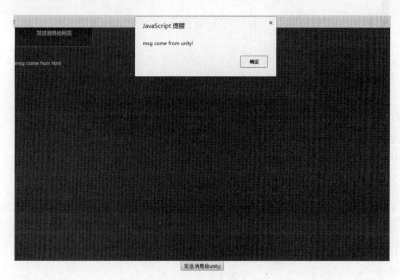

图 9-9　运行效果

9.1.3　Unity 发布为 Web 网页，在 Web Player 中打开一个新页面不被拦截

可以在需要打开网页的时候，先隐藏 WebPlayer，然后将网址传给超链接并且将超链接显示出来。最后单击超链接的时候就不会被浏览器拦截了。

在 Unity 中创建一个名为"OpenHtml.cs"的脚本，代码如下：

```csharp
using UnityEngine;
using System.Collections;

public class OpenHtml : MonoBehaviour
{
    private string stringToEdit = "";
    void Start ()
    {
        Application.ExternalCall("hide");
    }

    void OnGUI ()
    {
        stringToEdit = GUI.TextField(new Rect(10,10,300,20),stringToEdit,25);
        if(GUI.Button(new Rect(100,100,200,30),"Hide the Unity webplayer"))
        {
            Application.ExternalCall("setURL",stringToEdit);
            Application.ExternalCall("hideWebPlayer");
        }
    }
}
```

hide()可以用来隐藏超链接；setURL 是用来将场景中输入的网址传递给网页中的超链接；hide

Unity 游戏开发实用教程

Web Player 用来隐藏 Web Player；将脚本拖到摄影机上，然后发布。发布完成以后，添加三个 js 函数，代码如下：

```
function setURL(url)
{
    document.getElementById("go").href = url;
}
Function hideWebPlayer()
{
    document.getElementById("UnityPlayer").style.visibility="hidden";
    document.getElementById("link").style.display="block";

}
Function hide()
{
    document.getElementById("UnityPlayer").style.visibility="visible";
    document.getElementById("link").style.display="none";

}
```

然后在 HTML 的 body 中添加两个控件：

```
<body>
    <p class="header"><span>Unity Web Player  | </span>OtherDemo</p>
    <dxxxxxxxxxxxxxxx>
        <div  id="link"  style="display:none">
        <input type="image" onclick=hide()   src="back.png"/>
        <a href="" id="go" target="_blank" ><img  border="0" src="go.png"/></a>
        <div>
        <div id= xxxxxxxxxx>
            <div class="mlssing">
                <a href=http://Unity3d.com/webplayer/ title="Unity Web Player.Install now!">
                    <img alt="Unity Web Player.Install now!" src="http://webplayer.Unity3d.com/installatior.
                </a>
            </div>
            <div class="broken">
                <a href="http://Unity3d.com/webplayer/ title="Unity Web Player. Install now! Restart your r
                    <img alt>="Unity Web Player. Install now! Restart your browser after install."
                        src=http://webplayer.Unity3d.com/installation/getUnityrestart.png width="193"height="
                </a>
            </div>
        </div>

    </div>
    <p class="footer">&laquo;created with <a href=http://Unity3d.com/Unity/ title="Go to Unity3d.com">Uni
<body>
```

下面看一下效果，打开网页如图 9-10 所示。

在输入框中输入 http://www.baidu.com，单击"Hide the Unity Web Player"按钮，会隐藏 Web Player，显示一个按钮一个超链接，如图 9-11 所示。

单击向右的箭头可以打开一个新网页，单击左边的箭头可以显示 Web Player，如图 9-12 所示。

第 9 章　常见问题、错误及插件介绍

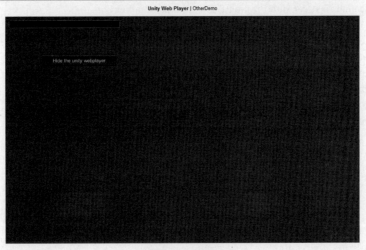

图 9-10　效果

图 9-11　超链接

图 9-12　完成效果

9.1.4　如何打开一个摄像头

通过 Unity 打开摄像头的代码如下:

```
using UnityEngine;
using System.Collections;

public class CameraTest : MonoBehaviour
{
    private WebCamTexture cameraTexture;
    private string cameraName="";
    private bool isPlay = false;

    void Start()
    {
        StartCoroutine(GetCameraTexture());
    }

    IEnumerator GetCameraTexture()
    {
        //如果是管理员，获取打开摄像头权限
        yield return Application.RequestUserAuthorization(UserAuthorization.WebCam);
```

```
            if (Application.HasUserAuthorization(UserAuthorization.WebCam))
            {
                    //获取第一个摄像头设备图像
                    WebCamDevice[] devices = WebCamTexture.devices;
                    cameraName = devices[0].name;
                    cameraTexture = new WebCamTexture(cameraName, 400, 300, 15);
                    Debug.Log(cameraName);
                    cameraTexture.Play();
                    isPlay = true;
            }
    }
    void OnGUI()
    {
        if (isPlay)
        {
    GUI.DrawTexture(new Rect(0, 0, 500, 350), cameraTexture, ScaleMode.ScaleToFit);
        }
    }
}
```

9.1.5 鼠标选中物体高亮

先在场景中创建 3 个 Cube，这 3 个 Cube 只有一个 Material，如图 9-13 所示。

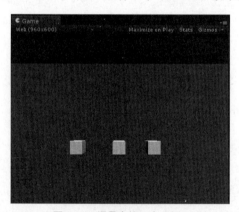

图 9-13 场景中的 3 个 Cube

下面单击 Cube 让它高亮，并且将上次单击的 Cube 颜色恢复，代码如下：

```
using UnityEngine;
using System.Collections;

public class HighLightDemo : MonoBehaviour
{

        //选中物体高亮
        //鼠标单击选中的物体
        private GameObject hitObj;

        //保存选中的物体颜色
```

```
    private Color savedColor;


    // Update is called once per frame
    void Update ()
    {
        //单击鼠标左键
        if(Input.GetMouseButtonDown(0))
        {
            CollisionDetection();
        }
    }

    void CollisionDetection()
    {

        Ray ray = Camera.main.ScreenPointToRay(Input.mousePosition);
        RaycastHit hit;
        if (Physics.Raycast(ray, out hit, Mathf.Infinity))
        {
            //假如从鼠标位置发射的射线从来没有与其他物体发生碰撞
            if(hitObj == null)
            {
                //保存本次碰撞物体信息方便下次恢复
                hitObj = hit.transform.gameObject;
                savedColor = hitObj.renderer.material.color;

                //改变物体颜色
                hitObj.renderer.material.color = Color.red;

            }
            else
            {
                //碰撞的不是同一个物体
                if(!hitObj.Equals(hit.transform.gameObject))
                {
                    //恢复上次物体的颜色
                    hitObj.renderer.material.color = savedColor;

                    //保存本次碰撞物体信息方便下次恢复
                    hitObj = hit.transform.gameObject;
                    savedColor = hitObj.renderer.material.color;

                    //改变物体颜色
                    hitObj.renderer.material.color = Color.red;

                }
            }
        }
        else
        {
```

```
            //没有射线碰撞的时候
            if(hitObj!=null)
            {
                //恢复上一次碰撞物体颜色
                hitObj.renderer.material.color = savedColor;
                hitObj = null;
            }
        }
    }
}
```

这个脚本需要注意的是场景中的物体都是单个材质,当遇到的物体是多个材质时,就需要用一个数组或者其他数据结构来存储单个物体上的材质。其次,需要了解 Physics.Raycast 这个函数,下面是该函数的参数列表:

Physics.Raycast 光线投射
static function Raycast (orinin:Vector3,direction:Vector3,distance:float+Mathf.Infinity,layerMask: int=kDefaultRaycastLayers):boll

最后一个参数用来设置物体的 Layer 层,关于 Layer 层可以看第 1 章,缺省为默认层。执行效果如图 9-14 所示。

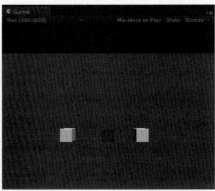

图 9-14 完成效果

9.1.6 如何打开一个本地 EXE

本例为打开本地的谷歌浏览器,脚本如下:

```
using UnityEngine;
using System.Collections;
using System.Diagnostics;
public class StartLocalExe : MonoBehaviour {

    void Start ()
    {
```

```
        //Process p = new Process();
        Process.Start(@"C:\Program Files (x86)\Google\Chrome\Application\chrome.exe");
    }
}
```

9.2 常见错误及解决

9.2.1 在使用 Unity 编写脚本时遇到的错误

当看到如图 9-15 所示的错误提示时,表下脚本名称和脚本中的类名不一致,将脚本名与类名统一即可。如图 9-16 所示为错误示例。

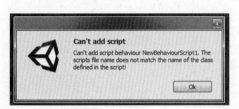

图 9-15　错误提示

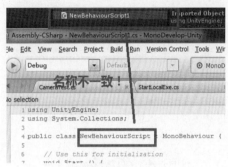

图 9-16　脚本名和类名不一致

9.2.2 使用 WWW 崩溃如何解决

WWW 正在加载资源的时候,如果调用了 Resources.UnloadUnusedAssets,就很容易导致场景崩溃,可以使用下面一句代码代替:

yield return Resources.UnloadUnusedAssets();

9.2.3 涉及 Direct3D 11 特效有时候不能显示出效果

例如,一个 Unity 的 Glass Refraction 材质包(玻璃材质),发布之后材质一开始没有显示,只有当单击 EXE 的边框时,材质才显示出来。

问题效果如图 9-17 至图 9-18 所示。

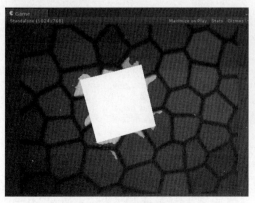

图 9-17　在 Unity 里面显示的效果

图 9-18　发布后打开效果

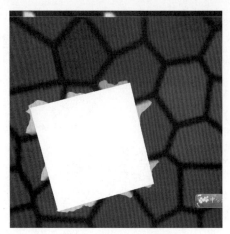

图 9-19 单击边框后显示

上面这种情况特别容易出现，尤其是 EXE 在 XP 的画面尺寸接近屏幕尺寸时（XP 其实不支持 DX11）。遇到这种问题可以用下面的方法解决：

（1）勾选 Player 设置里面的 "use Direct3D 11"。
（2）重启一下 Unity。
（3）发布。

9.2.4 引用 dll 的时候报错

这种情况可能是因为从网上下载的 dll 文件有问题，比如可能出现下面的错误提示：

VerificationException:Error verifying I18N.Common.ByteEncoding:GetString (byte[],int,int)

见到类似的错误提示时，解决办法就是去 Unity 安装目录下找 dll 文件，替换掉原来 Plugins 的就可以了。一般路径为 "...\Unity\Editor\Data\Mono\lib\mono\Unity"。

9.2.5 读取 Xml 错误

错误提示为："XmlException: Text node cannot appear in this state"，这个问题是 utf-8 编码的问题，如果是 utf-8+bom，那么就会出现这种问题，如果是单纯的 utf-8 就没有这种问题。可以手动改，也可以修改读取代码，代码如下：

```
public bool LoadXml(string xmlFile)
{
        System.IO.StringReader stringReader = new System.IO.StringReader(xmlFile);
        stringReader.Read(); // 跳过 BOM
        System.Xml.XmlReader reader = System.Xml.XmlReader.Create(stringReader);
        xmlDocument.LoadXml(stringReader.ReadToEnd());
    return true;
}
```

9.2.6 Fail to download data file

当发布为网页并部署到 IIS 上的时候，可能出现如图 9-20 所示的问题。

这是因为 Web 服务器并没有支持*.Unity3d 的文件类型。需要在 Web 服务器中添加 MIME 类型，如图 9-21 所示，双击 MIME 类型。

图 9-20　错误提示　　　　　　　　图 9-21　IIS 面板

然后添加.Unity3d 和 application/octet-stream，如图 9-22 所示

图 9-22　添加 MIME 类型

如果是 IIS6，则鼠标右键单击计算机名，选择属性，找到 MIME 类型，然后添加即可。

9.3　Unity 插件

虽然本书对插件的介绍篇幅并不多，但是插件在实际的开发中往往有着难以忽视的作用。它来源于 Unity，但却并非完全属于 Unity。下面列出了部分插件：

（1）界面插件：NGUI、igui。

（2）足球插件：Soccertoolkit。

（3）可视化脚本工具：Playmaker。

（4）切西瓜插件：Shatter toolkit。

（5）触摸插件：FingerGestures 、EasyTouch。

（6）车辆游戏插件：Edy's Vehicle Physics。

（7）天空插件，模拟各种天气：Unisky。

（8）字体插件：GlyphDesigner。

（9）3d 红绿插件：3d anaglyph system。

（10）寻路插件：EasyRoads3Dv1.8.1.Unity package。

（11）游戏记录回放插件：EZ Replay Manager。

（12）动画插件：iTween-Visual-Editor。

（13）移动平台视频纹理：Mobile Movie Texture。
（14）魔幻特效插件：MagicalEffects.Unity package。
（15）外发光着色器插件：Glow 11 Unity3d。
（16）向量线工具：Vectrosity。
（17）着色器包插件：Aubergines Shader Pack。
（18）体感插件：KinectExtras with MsSDK。
（19）虚拟游戏摇杆：EasyJoystick。
（20）增强实境工具：Qualcomm AR。
（21）制作特效的工具：FX Maker。
（22）液体、水、浮力模拟开发：Buoyancy Toolkit v1.41。
（23）声音特效工具包：Explosion Sound Effects。
（24）跑酷插件：Ultimate Endless Runner Kit v1.03。
（25）变形插件：Megafiers.Unitypackage。
（26）相机控制器插件：(MMO)RPG Camera & Controller。
（27）破碎特效插件：Chipoff-fracture-system。
（28）摄影机漫游路径动画插件：Camera Path Animator。
（29）粒子效果插件：Particle-system-collection。
（30）地形资源插件：TerrainToolkit。
（31）广告插件：Prime31_StoreKit_for_iPhone。
（32）镜头特效插件：Camera splat effects。
（33）背包系统插件：InventoryManager-js-cs。
（34）材质编辑插件：Strumpyshadereditor。

第 10 章　多人在线的坦克大战

本章将本书的一些知识点串联起来，制作一个多人在线的坦克大战游戏。主要功能点有：二维界面制作、地形制作、粒子特效、数据库登录验证、保存信息到数据库、Xml 的读取、声音播放、多人在线坦克的运动及攻击和计分。

10.1　项目介绍

10.1.1　游戏主要功能描述

（1）界面部分：主要为登录、注册界面制作。
（2）地形部分：主要为登录场景以及游戏场景地形。
（3）粒子特效：火焰粒子特效、爆炸粒子特效、烟雾粒子特效。
（4）登录验证：数据库的读取与写入。
（5）音效：开炮音效、背景音乐、爆炸音效。
（6）坦克部分：主要为坦克的前进、后退、左转、右转、子弹发射、碰撞检测。
（7）多人在线部分：多玩家运动同步、子弹伤害、玩家销毁、计分。

10.1.2　游戏开发步骤介绍

首先搭建场景，包括创建地形、坦克模型设置以及粒子特效；然后分场景进行功能开发。登录场景主要为登录验证和注册，游戏场景主要为多玩家坦克的运动控制、攻击以及小地图。

游戏的思路：在 Start 场景中，当一个玩家注册时首先需要选择所在阵营，是红方还是蓝方，注册完以后，登录系统会获取玩家信息，包括用户名、密码、分数、阵营。一旦验证登录成功就会进入 Game 场景。

进入 Game 场景时玩家会连接服务器，连接成功后会立即生成所有在线玩家。在实例化玩家的时候会自动根据玩家的阵营信息来生成不同种类的坦克，同时也会修改所有玩家的 tankinfo 信息。当然在实例化玩家的时候会随机设置玩家的位置，并且显示本地玩家的分数。

玩家实例化的时候，非本地玩家会关闭一些运动控制代码（Player.cs），这样本地玩家就可以控制本地玩家，这样就不会在控制本地玩家的时候其他坦克也在运动。同时玩家的阵营也根据坦克身上的炮弹的 Tag 值来设定。

Network View 这个组件的存在可以让所有玩家位移、旋转同步到其他客户端，但是发射炮弹、受到伤害、计分等这些功能还需要通过 RPC 调用来实现。

Unity 的碰撞检测在子弹运动的过程中并非每次都能检测得到，所以使用了射线碰撞检测来检测碰撞。

当不同的坦克相遇的时候，一个坦克发射了一枚炮弹打中另一个坦克，这个时候炮弹上面的脚本就会判断：（1）如果是同一阵营坦克，不扣生命数，否则被击中坦克扣一命；（2）如果炮弹打中的是本地玩家，判断是否是同一阵营，不是的话本地玩家要扣一条命；（3）如果炮弹是本地玩家发射的，并且被击中坦克的生命值小于 0，那么销毁击中坦克，本地玩家加一分，击中的坦克向所有客户端广播被击毁了。

有时候我们很难判断炮弹究竟是本地玩家还是其他客户端发射的，所以本地玩家发射的时候会将本地玩家的信息保存到发射的子弹上。实际上本游戏在判断交战情况时并非完全同步。这里使用了本地提前预判。在本地一个玩家被消灭时，被销毁的玩家客户端会收到被消灭的信息。所以这个地方有可能不同步。

如果玩家被消灭了，我们应当允许玩家重新开始，所以当他单击重新开始的时候会自动重新加载 Game 场景，这样他就可以重新生成了。

当我们单击 Esc 退出时，会弹出一个按钮退出，退出的同时玩家的最新分数会上传到数据库中，下一次登录的时候会读取新的数据。

10.2 前期准备以及场景搭建

10.2.1 前期准备

新建一个名为"TankGame"的工程，在 Project 面板下建立如图 10-1 所示的目录。

Data 文件夹用来存放可能需要引用的外部资源文件；Plugins 存放一些 dll 文件；TankGame 文件夹存放游戏的主要资源以及脚本文件，其中 Auido 存放声音文件、Models 存放一些模型和预设文件、Scenes 存放场景、Scripts 存放脚本文件、Terrain 存放地形文件。

下面将模型文件以及声音文件导入到对应的文件夹中，如图 10-2 所示。

图 10-1　文件目录结构

图 10-2　资源文件

10.2.2 搭建场景

新建两个场景，第一个场景名为"Start"，第二个场景名为"Game"，然后单击进入"Start"场景，新建一个 Terrain，并且刷出一些树木和山峰，如图 10-3 所示为完成效果。

Start 场景以及 Game 场景的地形效果具体制作过程请参考第 6 章第 2 节。效果可以自己去把握，这里使用了 Unity 自带的 2 个资源包，如图 10-4 所示。

图 10-3 创建地形效果

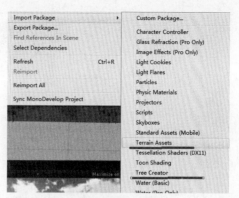
图 10-4 导入地形包

同样在 Game 场景中创建一个 Terrain，效果如图 10-5 所示。Game 场景中的地形最好大一些，Start 场景中的 Terrain 可以小一点。回到 Start 场景中，放入一些模型点缀场景，如图 10-6 所示。

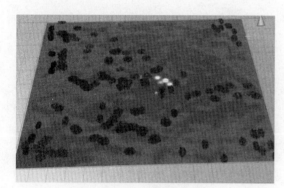
图 10-5 Game 场景中的地形

图 10-6 Start 场景

调整 Camera 的初始视角到一个合适的角度，如图 10-7 所示。

图 10-7 Start 场景

主体场景已经搭建完毕，但是场景看起来还是比较呆板，所以下面添加几个粒子特效使场景更加富有动感。导入 Unity 自带的资源文件包，如图 10-8 所示。

导入粒子特效包后，对其中的火焰粒子特效和爆炸粒子特效做简单修改并保存，如图 10-9 所示。

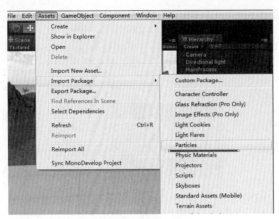

图 10-8 添加粒子特效包

图 10-9 完成的特效

下面将导入的坦克模型拖到场景中,重命名为 tank1 和 tank2,并且在位于坦克炮管的前方添加一个名为 firePoint 的空体物,这个空物体作为炮弹发射的起点。如图 10-10、图 10-11 所示。

图 10-10 坦克开火点 firePoint

图 10-11 开火点相对位置设置

10.2.3 设置游戏背景音乐

打开 Start 场景,在 Camera 上添加一个名为 Audio Source 的组件,将资源文件夹中的"BackgroundMusic"音效拖到 Audio Source 中的 Audio Clip 上,并设置音量 Volume 为 0.5,设置选项如图 10-12 所示。

同样打开 Game 场景,给摄影机添加 Audio Source 组件,将音量设置小一点就可以了,设置如图 10-13 所示。

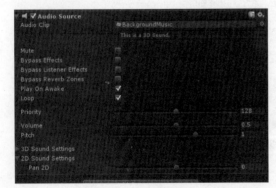

图 10-12 添加背景音乐

图 10-13 Game 场景中背景音乐

10.3 登录场景开发

10.3.1 登录场景界面制作

登录场景界面主要分为初始界面、登录界面、注册界面。其中注册界面中需要做一个下拉菜单。初始界面主要为一个窗口，其中有三个按钮，上方有"坦克大战"四个字，如图 10-14 所示。

这里主要用 GUISkin 中的风格以及部分自定义的 GUIStyle 来制作界面。GUISkin 可以对一些通用的控件风格做统一设置。将这个自定义的 GUISkin 放在 Resources 文件夹中，如图 10-15 所示。

图 10-14 登录效果

图 10-15 Skin 使用

如果需要使用 Skin 中的风格，可以使用以下代码导入：

```
//皮肤
    private GUISkin skin;.....
void Start ()
{
skin = Resources.Load("Skin/SciFi Buttons GUI") as GUISkin;

}
void OnGUI ()
{
GUI.skin = skin;
}
```

下面为初始界面的主要代码：

```
GUI.Label(new Rect(760,30,160,60),"坦克大战",labelStyle);
    GUI.skin = skin;
    GUI.Box(new Rect(800,150,420,350),"");
    if(GUI.Button(new Rect(850,230,300,50),"用户注册"))
    {
        isOpenRegisterWindow = true;
    }
    if(GUI.Button(new Rect(850,300,300,50),"用户登录"))
    {
```

```
                isOpenLoginWindow = true;
            }


            if(GUI.Button(new Rect(850,370,300,50),"游戏退出"))
            {
                Application.Quit();
            }
```

单击"用户注册"会显示如图 10-16 所示的界面。

下面介绍在 Unity 中制作下拉菜单的方法。Unity 的界面系统控件中是没有现成的下拉菜单控件的，但是可以通过一些控件间的组合来制作一个下拉菜单。原理如下：

当我们单击"红队"这个按钮（没错，这个看起来像输入框的控件实际上是一个按钮）就显示一个窗体，窗体内有两个按钮，按钮铺满窗口。同时程序要判断鼠标是否在下拉按钮及窗体内，如果在，那么显示窗体，否则不显示窗体；当鼠标在下拉列表范围内时，单击窗体内的按钮，则修改所属阵营并且隐藏窗体。如图 10-17 所示为下拉列表显示情况。

图 10-16 注册界面

图 10-17 下拉窗口

下拉列表主要代码如下：

```
//下拉窗口，如果在下拉窗口范围内，窗口显示，否则关闭
        if(isClickChooseTeamButton)
        {
            chooseTeamWindowRect = GUI.Window(0, chooseTeamWindowRect, DoTeamTypeWindow,
"",chooseTeamWindowStyle);
            float mouse_x = Input.mousePosition.x/Screen.width*1366;
            float mouse_y = (Screen.height - Input.mousePosition.y)/Screen.height*768;
            if(showChooseTeamWindowRect.Contains(new Vector2(mouse_x,mouse_y)))
            {
                isClickChooseTeamButton = true;
            }
            else
            {
                isClickChooseTeamButton = false;
            }

        }
```

这里使用了自适应代码,所以在任何屏幕尺寸下,都可以把屏幕尺寸看作是1366×768,那么鼠标在屏幕的位置就要重新换算,代码如下所示:

float mouse_x = Input.mousePosition.x/Screen.width*1366;
float mouse_y = (Screen.height - Input.mousePosition.y)/Screen.height*768;

登录界面就比较简单了,可以参考工程文件自己理解,效果如图10-18所示。

图10-18 登录界面

10.3.2 玩家注册功能

界面完成以后,首先搭建MySQL+Apache环境。安装好Xampp后打开MySQL数据库,手动新建一个数据库,名为"Tank",并新建一张名为"tankusers"的表,如图10-19所示。

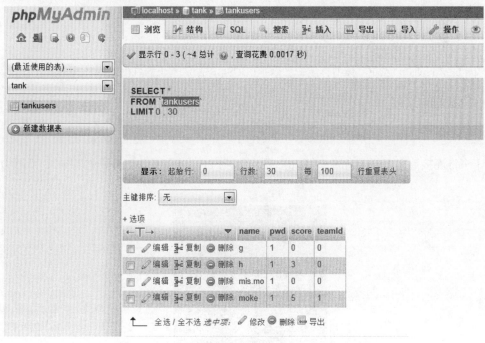

图10-19 数据库

将Connect.php这个脚本放到Xampp安装目录下的htdocs文件夹中,如图10-20所示。

图 10-20　网站下的 Connect.php

Connect.php 代码如下：

```php
<?php
    $db=mysql_connect('localhost','root','qwer1234') or die('Unable to connect:'.mysql_error());
    mysql_query("SET NAMES 'utf8'");
    $b = mysql_select_db('tank',$db) or die('Unable to connect to the database');
    $sql = $_POST['sql'];
    $result = mysql_query($sql) or die('Query failed!'.mysql_error());
    $info[]=array();
    $i=0;
    while($rs = mysql_fetch_assoc($result))
    {
     $info[$i]=$rs;
     $i++;
    }
    $infostr=json_encode($info);
    $infostr= preg_replace("#\\\u([0-9a-f]{4})#ie", "iconv('UCS-2BE', 'UTF-8', pack('H4', '\\1'))", $infostr);
    echo $infostr;
    mysql_close($db);
?>
```

接下来将连接数据库要用到的一些 dll 文件导入到 Project 下的 Plugins 文件夹中，如图 10-21 所示。

图 10-21　调用数据库要用到的 dll

下面开始编写代码。注册主要代码如下：

代码片段一：

```
void DoRegisterWindow(int windowID)
    {
        GUI.Label(new Rect(200,5,100,50),"注 册");
        GUI.Label(new Rect(30,70,200,50),"用户名: ");
        registerName = GUI.TextField (new Rect(210, 70, 230, 35), registerName);
        GUI.Label(new Rect(30,140,200,50),"密 码: ");
        registerPwd = GUI.TextField (new Rect(210, 140, 230, 35), registerPwd);
         GUI.Label(new Rect(30,210,200,50),"阵 营：");
        if(GUI.Button(new Rect(210,210,230,40),teamName))
        {
             isClickChooseTeamButton = true;
        }
        if(GUI.Button(new Rect(130,280,100,40),"注 册"))
        {
             if(registerName.Length>0 && registerPwd.Length>0)
             {
                 InsertInfoToDB(registerName,registerPwd,Connect.typeId.ToString());
                 isOpenRegisterWindow = false;
             }
            else if(registerName.Length == 0)
            {
                errorMsg = "注册名不能为空";
            }
            else if(registerPwd.Length == 0)
            {
                errorMsg = "密码不能为空";
            }


        }
        if(GUI.Button(new Rect(250,280,100,40),"取 消"))
        {
            isOpenRegisterWindow = false;



        }
    }
```

代码片段二：

```
void InsertInfoToDB(string name,string pwd,string teamId)
{
    string sql = "insert into tankUsers values('!!!','@@@','&&&','###')";
    sql= sql.Replace("!!!",name).Replace("@@@",pwd).Replace("###",teamId).Replace("&&&","0");
    StartCoroutine(OperateDatabase.LoadData(sql));
```

_Unity 实用开发教程

}

代码片段三：

```
using UnityEngine;
using System.Collections;

public class OperateDatabase : MonoBehaviour {

    public static string conUrl = "http://127.0.0.1/connect.php";
    //从数据库获取的结果
    public static string result;
    void Start ()
    {
        result = "";
    }
    public static IEnumerator LoadData(string sql)
    {

        WWWForm form = new WWWForm();
        //提交 sql 语句
        form.AddField("sql",sql);
        //向服务器提交需求
        WWW hspost = new WWW(conUrl,form);
        yield return hspost;
        if(hspost.error != null)
        {
                print(hspost.error);
        }
        else
        {
                result = hspost.text;

        }
    }

}
```

10.3.3　数据库登录验证

登录验证的时候主要验证用户名和密码是否正确，代码如下：

```
IEnumerator UserAuthentication(string name,string pwd)
{
    string sql = "select * from tankUsers where name='&&&'";
    sql = sql.Replace("&&&",name);
    yield return StartCoroutine(OperateDatabase.LoadData(sql));
    if(OperateDatabase.result.Length>0)
    {
```

```
            JsonData[]usersInfo =JsonMapper.ToObject<JsonData[]>(OperateDatabase.result);

            if(usersInfo.Length == 1)
            {
                if(usersInfo[0]["pwd"].Equals(pwd))
                {
                    user.name = name;
                    user.pwd = pwd;
                    user.teamId = usersInfo[0]["teamId"].ToString();
                    user.score = usersInfo[0]["score"].ToString();
                    Connect.typeId = int.Parse(user.teamId);
                    Application.LoadLevel("Game");
                    Debug.Log("登录成功");
                }
                else
                {
                    errorMsg = "密码错误！";
                }
            }
            else if(usersInfo.Length == 0)
            {
                errorMsg = "用户不存在！";
            }


}
```

用户登录和用户注册都会用到一个名为 OperateDatabase 的类，其中使用 WWWForm 来生成表单数据，然后通过 WWW 来上传。代码如下：

```
Public static string conUrl = "http://127.0.0.1/connect.ptp";
//向服务器提交需求
WWW hspost = new WWW(conUrl, form);
```

打开 OperateDatabase.cs 这个脚本，特别要注意由于测试的时候是在本地测试，使用的 IP 为 127.0.0.1，读者在使用该脚本时需要修改该 IP 地址为服务器网络地址，否则会报错。

10.4 游戏场景开发

10.4.1 创建一个服务器

新建一个名为 Server.cs 的脚本，创建服务器主要脚本，代码如下：

```
if(GUI.Button(new Rect(10,10,100,60),"创建服务器"))
{
        Network.InitializeServer(16, connectPort, false);
}
```

在多人在线服务器脚本中还有一个函数必须加上，这个函数主要用来处理当用户从服务器断开

Unity 实用开发教程

时服务器的一些操作。

```
void OnPlayerDisconnected(NetworkPlayer player)
{
    Network.RemoveRPCs(player);
    Network.DestroyPlayerObjects(player);
}
```

10.4.2 多人在线坦克行为模块开发

坦克的行为模块包括位移运动以及发射炮弹。通过控制 W、S、A、D 来控制坦克的前进、后退、左转、右转；通过按下 Space 键来发射炮弹。坦克的行为模块代码如下：

```
if (Input.GetKey(KeyCode.W))
{
    transform.Translate(Vector3.forward * Time.deltaTime * movespeed);
}
if (Input.GetKey(KeyCode.S))
{
    transform.Translate(Vector3.forward * Time.deltaTime * (-movespeed));
}
if (Input.GetKey(KeyCode.D))
{
    transform.Rotate(Vector3.up * Time.deltaTime * rotatespeed);
}
if (Input.GetKey(KeyCode.A))
{
    transform.Rotate(Vector3.up * Time.deltaTime * (-rotatespeed));
}
if (Input.GetKeyDown("space"))
{
//发射部分代码请看原工程文件
if(myInfo.bullitCount>0)
    {
        myInfo.bullitCount --;
        bullit.GetComponent<Bullit>().myOwner = gameObject;
        Instantiate(bullit, firePoint.position, firePoint.rotation);
        Instantiate(Fire,firePoint.position,firePoint.rotation);
        gameObject.networkView.RPC("InstantiateBullit",RPCMode.Others,firePoint.position,firePoint.rotation);

    }
    else
    {
        Debug.Log("弹药耗尽!");
    }
}
```

对于发射子弹代码，首先要判断自己是否还有炮弹，如果有的话，那么就实例化一个炮弹并且同时实例化一个开炮效果的粒子特效，同时调用其他客户端的这台坦克发射炮弹，发射完后子弹数量减一。

10.4.3 登录后自动连接服务器并生成玩家

Connet.cs 主要是用来连接服务器，代码如下：

```
Private-string connectToIP;
    private    int connectPort;
    public static int typeId = 0;
    void Start ()
    {
        ReadXmlFile readXml = new ReadXmlFile();
        readXml.Load("file:///"+Application.dataPath+"/Data/Xml/ServerInfo.xml");
        connectToIP = readXml.GetRootNode().SelectSingleNode("ip").InnerText;
        connectPort= int.Parse(readXml.GetRootNode().SelectSingleNode("port").InnerText);
        Network.useNat = false;
        Network.Connect(connectToIP, connectPort);
    }
```

这里需要将服务器的 IP 与端口号写在 Xml 中，所以在连接服务器之前需要先从 Xml 中获取服务器 IP 和端口号。

10.4.4 炮弹的功能开发以及记分

如图 10-22 所示为炮弹的设置界面。

图 10-22　炮弹设置页面

炮弹上需要加上一个音效，该音效自动播放，同时需要给它加上 Bullit.cs 脚本，注意由于炮弹与坦克发生碰撞检测时有时候会出现问题（这段代码已经被注释），所以这里使用了射线碰撞检测。当与本地玩家（Player）发生碰撞的时候，判断炮弹是否来自敌方阵营，如果来自敌方阵营那么生命值减一，当本地玩家生命值为 0 的时候销毁本地玩家。如果是其他坦克，如果是不同阵营生命值减一，如果子弹来自本地玩家并且被攻击的坦克生命值为 0，本地玩家加一分，并且子弹数目为 10。

炮弹的 Tag 值被特别设置，Tag 值为 0 或者为 1。0 代表红方，1 代表蓝方。在玩家坦克初始化的时候会根据 Tag 值来设置坦克阵营。

10.4.5 多人在线游戏小地图开发

在这个游戏中制作了一个简单的小地图,它能够显示所有玩家的位置。本地玩家为黄色点;红队为红点;蓝队为蓝点。如图 10-23 所示。

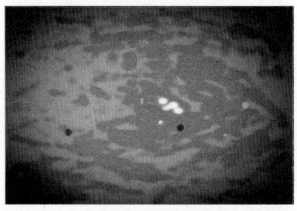

图 10-23　小地图完成效果

小地图的制作方法在第 2 章已经介绍过,这里不再重复。多人在线的小地图的难点在于需要将其他玩家坦克的位置标记在小地图上。对于这个问题,可以在 UIManager.cs 脚本中建立 2 个 list,代码如下:

```
public List<Transform> blueTeams;
public List<Transform> redTeams;
```

当坦克实例化以后,绑定在坦克上面的 TankInfo.cs 就执行 Start()这个函数,在 Start()中执行添加到队列中去的操作,代码如下:

```
void Start ()
    {
        uiManager = GameObject.Find("MainProcess").GetComponent<UIManager>();
        lives = 3;
        if(gameObject.name.Equals("Player"))
        scores = int.Parse(StartUI.user.score);
        bullitCount = 10;
        if(typeId == 0)
        {
//本地玩家不放入队列中
            if(!gameObject.tag.Equals("Player"))
            {
                uiManager.redTeams.Add(gameObject.transform);
            }

            Debug.Log("红");
        }
        else
        {
            if(!gameObject.tag.Equals("Player"))
```

```
            {
                uiManager.blueTeams.Add(gameObject.transform);
            }
            Debug.Log("蓝");
        }
    }
}
```

这样就可以获取所有在线玩家的位置信息了。

当玩家销毁的时候，就必须要从相应的队列中移除玩家的 Transform 组件。这个操作可以放在 Bullit.cs 中。代码如下：

```
if(tankInfo.typeId == 0)
{
uiManager.redTeams.Remove(gameObject.transform);
}
else
{
    uiManager.blueTeams.Remove(gameObject.transform);
}
```

10.4.6 退出游戏并提交成绩到数据库

当单击 Esc 的时候会弹出如图 10-24 所示的窗口。

图 10-24 退出菜单

当单击退出时，先判断成绩有没有改变，如果有改变，就将本地成绩上传到数据库，然后退出，代码如下：

```
if(GUI.Button(new Rect(200,30,140,50),"退 出"))
{
    if(myInfo!=null)
    {

        if(!StartUI.user.score.Equals(myInfo.scores.ToString()))
        {

StartCoroutine(SaveAndExit(StartUI.user.name,myInfo.scores.ToString()));
        }
        else
        {
            Application.Quit();
        }
    }
}
```

SaveAndExit 函数代码如下：

```
IEnumerator SaveAndExit(string name,string score)
{
    string sql = "update tankUsers set score='&&&' where name='$$$'";
    sql = sql.Replace("&&&",score).Replace("$$$",name);
    yield return StartCoroutine(OperateDatabase.LoadData(sql));
    Application.Quit();

}
```